Journal Edition
for Weekdays and Shabbat

משכן תפלה
MISHKAN T'FILAH

A Reform Siddur

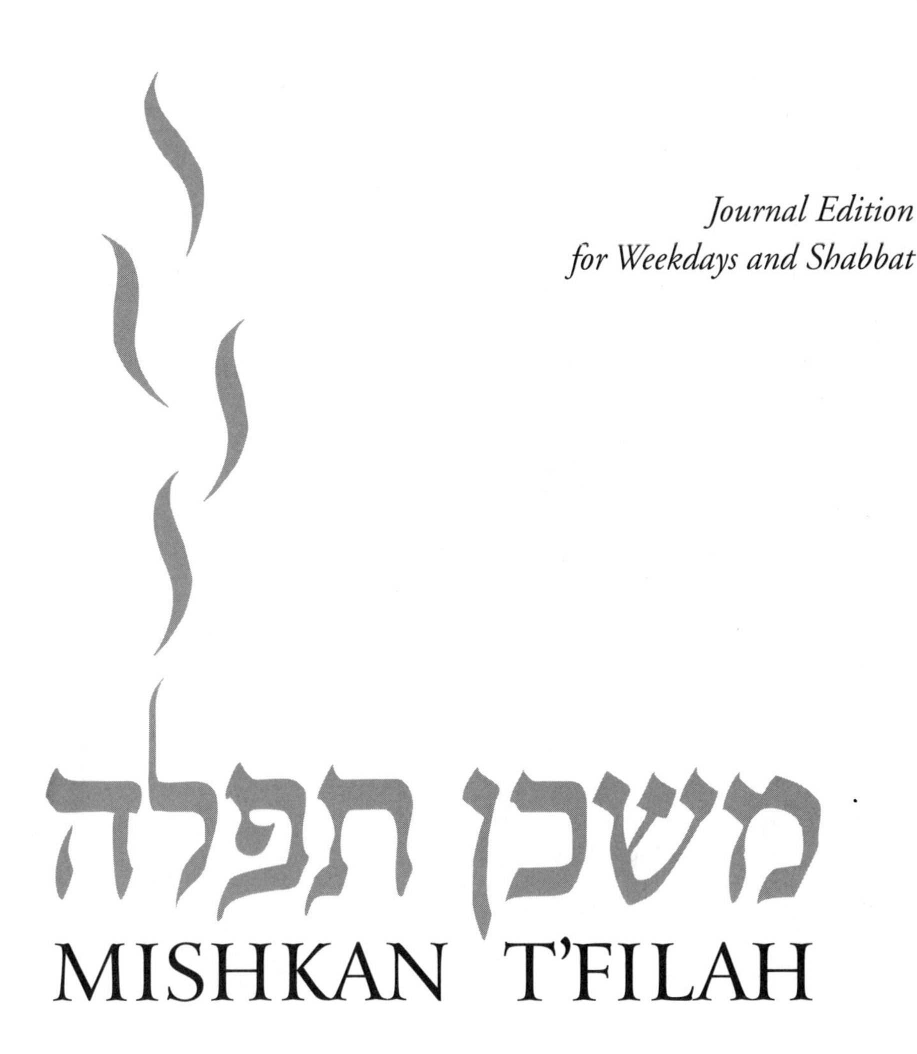

A Reform Siddur

וְעָשׂוּ לִי מִקְדָּשׁ וְשָׁכַנְתִּי בְּתוֹכָם

"And let them build Me a sanctuary
that I may dwell in their midst."

CENTRAL CONFERENCE OF AMERICAN RABBIS
5771 NEW YORK 2010

Note: Page numbers in this special edition are meant to be consistent with the page numbering of the standard, complete edition of *Mishkan T'filah.* To achieve this goal, there are several jumps in page numbering in this edition.

15 14 13 12 7 6 5 4 3 2

ISBN 978-0-88123-124-3

CCAR Press, 355 Lexington Avenue, New York, NY 10017
(212) 972-3636
www.ccarpress.org

Original Edition: Edited by Elyse D. Frishman
Journal Edition: Edited by Joel Abraham and Michelle Shapiro Abraham

Designed by Neil Waldman. Typography by Nostradamus Advertising.
Cover design by Izzy Pludwinski.

Contents

MISHKAN T'FILAH JOURNAL EDITION TASK FORCE

Joel N. Abraham and Michelle Shapiro Abraham
Hara E. Person

Lisa Barzilai	Paul Kipnes
Ana Bonnheim	Avram Mandell
Jennifer Clayman	Rob Nosanchuk
Stephanie Fink	Stacy Rigler
Richard Kellner	P.J. Schwartz

Steven A. Fox, *ex officio*

How Do I Use This Book?

Most prayers in *Mishkan T'filah* are presented on a two-page spread. The right-hand page has the prayer in Hebrew, transliteration, and a modern, faithful translation into English. In the original version of *Mishkan T'filah,* the left-hand page voices the prayer through the words of others — modern poets, liturgists, theologians, and others. In *Mishkan T'filah: The Journal Edition,* the left-hand side is blank, to leave room for **your** voice. Read the prayer on the right-hand side and then respond to its words, its rhythms, or its themes. Focus on part of the prayer or react to all of it. Use the last line (or *chatimah*) as a starting point or, like in the original *Mishkan T'filah,* an ending point. On the left-hand side, write a poem; draw a picture. Share your own feelings or experiences ... or try writing a prayer.

Some left-hand side pages have questions to help bring you in or to challenge you to see the prayer differently. These questions contain one of four possible different approaches (explained below). Answer the questions, or respond to them with more questions. Pick a different type of question for each prayer, or take a journey through one type of question all the way through the prayerbook. (Additional questions can be found on the Web site www.myMTJ.org, along with other ways to bring yourself into the *siddur.*) The questions were created based on these four approaches:

(I) signifies an **Imagery** question. These are designed to help you imagine your way into the prayer. They can also be used for guided imagery exercises.

(E) indicates an **Experiential** question. These are designed to make a link between experience, memory and the theme of the prayer.

(T) is for a **Trigger Text** — usually a text external to the prayer, to compare and contrast with it.

(P) means **Paradigm Shift**. These questions are meant to turn the prayer around and give an alternate way to consider yourself and the prayer.

(For other ideas and ways to use *Mishkan T'filah: The Journal Edition,* check out www.myMTJ.org and click on "teaching.")

How Do I Pray With This Prayerbook?

Mishkan T'filah: The Journal Edition is designed to be used in conjunction with other versions of *Mishkan T'filah.* You can bring this prayerbook to services at your synagogue, your URJ camp, or NFTY Kallah or convention. The prayers on the right-hand side will be the same as the pages in the original *siddur* and can be used as normal. When participating in a service, you can record your thoughts on the left-hand side to help bring more meaning to your worship. When leading a service, sharing those thoughts can help bring meaning to others. Another exercise might be to assign different prayers to participants (or groups of participants) before a service. Encourage them to fill in the left-hand page and then, during the service, share their voices with the *kahal* (community).

For more detailed suggestions of how to pray with this prayerbook or for examples of what others have done with *Mishkan T'filah: The Journal Edition* as a whole or on various left-hand pages, go to www.myMTJ.org. Please join the community to share your voice as we create a virtual prayer space online.

Introduction to the Original Edition

While prayer invites us to beseech God, we must also be open to what God wants from us. Samuel Karff wrote, "Each generation must struggle to hear the call, 'Where art thou?' Each must choose to answer, 'Here I am, send me.'" *Each generation* — not merely each individual. A *siddur* must challenge narcissism; that challenge begins by saying to a worshipper: your voice is here amidst others. *To hear the call:* to realize that prayer is not merely an outpouring of self; it is the opening of our senses to what is beyond our selves. *Send me:* prayer must motivate us to give selflessly.

In any worship setting, people have diverse beliefs. The challenge of a single liturgy is to be not only multi-vocal, but poly-vocal — to invite full participation at once, without conflicting with the *keva* text. (First, the *keva* text must be one that is acceptable; hence, the ongoing adaptations of certain prayers, over time, such as the *G'vurot*). Jewish prayer invites interpretation; the left hand material was selected both for metaphor and theological diversity. The choices were informed by the themes of Reform Judaism and Life: social justice, feminism, Zionism, distinctiveness, human challenges. The heritage of Reform brings gems from the *Union Prayer Book* and from *Gates of Prayer,* as well as from Reform's great literary figures over the last century and more.

Theologically, the liturgy needs to include many perceptions of God: the transcendent, the naturalist, the mysterious, the partner, the evolving God. In any given module of prayer, e.g., the *Sh'ma and Its Blessings,* we should sense all of these ways. The distinction of an integrated theology is not that one looks to each page to find one's particular voice, but that over the course of praying, many voices are heard, and ultimately come together as one. The ethic of inclusivity means awareness of and obligation to others rather than mere self-fulfillment.

An integrated theology communicates that the community is greater than the sum of its parts. While individuals matter deeply, particularly in the sense of our emotional and spiritual needs and in the certainty that we are not invisible, that security should be a stepping stone to the higher value of community, privilege and obligation. We join together in prayer because together, we are stronger and more apt to commit to the values of our heritage. Abraham knew that just ten people made a difference. In worship, all should be reminded of the social imperatives of community.

Prayer must move us beyond ourselves. Prayer should not reflect "me"; prayer should reflect *our* values and ideals. God is not in our image; we are in God's. It is critical that Reform Jews understand what is expected of them. The diverse theologies of the new *siddur* include religious naturalism, the theology of human adequacy, process theology, and the balance of particularism and universalism. But the essence of Reform liturgy continues to be what God demands of us, with heavy emphasis on ethical action and social justice.

In *Beyond the Worship Wars,* Thomas G. Long teaches, "Part of the joy of worship is to know the motions, know the words, know the song. The vital congregations knew their order of worship and moved through it with deep familiarity. What is more, the worshippers had active roles — speaking, singing, moving — and many of these they could perform from memory." The *siddur* is a tool in the larger system of worship. Lawrence Hoffman teaches, "The book is less text than pre-text for the staging of an experience. We are returning to the age of orality, where performance of prayer matters more than the fixed words. The question of worship leadership has expanded now, to include the theology and artistry of being a *sh'liach tzibur* — how to orchestrate seating, fill empty space, provide the right acoustics, and honor individualism within the group experience."

Using *Mishkan T'filah,* the actual selection of prayer can wait for the moment. The *sh'liach tzibur* must offer a recipe that works comfortably for the community, and be able to adapt each week to the particular needs of the community, and to individuals within that community.

Mishkan T'filah invites familiarity, even as it allows for diversity. Over time, one cannot help but memorize the book. The content of each page spread, though varied, becomes known. The constancy of the *keva* text (the right hand side of each page which offers the traditional prayer) anchors every creative prayer on the left. It is the cumulative effect of worshipping from this *siddur* that will deepen meaningful ritual.

The publication of *Mishkan T'filah* continues the Reform movement's tradition of liturgical innovation. A single prayer book provides an important vehicle for group identification as well as personal prayer. *The Union Prayer Book* and its successor *Gates of Prayer* and now *Mishkan T'filah* each expresses the ethos and values of its own era, at the same time being fully rooted in the structure and substance of the historical liturgical tradition of the Jewish people.

The title *Mishkan T'filah* is drawn from Exodus 25:8 where God commands us to build a portable sanctuary that can accompany us on our wanderings. "And let them build Me a sanctuary that I may dwell among them." *Mishkan T'filah* is a dwelling place for prayer, one that moves with us wherever we might be physically or spiritually. It offers the opportunity for God, the individual and community to meet. The desert *mishkan* was a portable sanctuary. Its care was guarded by the Levites and the priests yet it invited all to bring their offerings. Today, we are all caretakers of *Mishkan T'filah*; may our offerings be acceptable before God.

May all who enter find joy, solace and meaning.

RABBI ELYSE D. FRISHMAN
Editor

RABBI PETER S. KNOBEL
Chair of the Editorial Committee

MISHKAN T'FILAH ORIGINAL EDITION

SIDDUR PUBLISHING TEAM [2004–2005]

Peter S. Knobel, Chair
Elyse D. Frishman, Editor

Lawrence A. Hoffman
Bernard H. Mehlman
Elliot L. Stevens
Elaine S. Zecher

Deborah Smilow, CCAR Managing Editor

EDITORIAL COMMITTEE [1999–2005]

Peter S. Knobel, Chair
Elyse D. Frishman, Editor
Elaine Zecher, Chair, Committee on Liturgy and Practices

Judith Abrams (consulting editor)
Roslyn Barak
William Cutter
Harry Danziger
Daniel Freelander
Lawrence Hoffman
Yoel Kahn
Lewis Kamrass
Jeffrey Klepper
Charles Kroloff
Richard Levy
Shelley Limmer
Janet Marder
George Markley
Yehoram Mazor
Paul Menitoff
Shira Milgrom
Richard Sarason
Benjie Ellen Schiller
Elliot Stevens
Sue Ann Wasserman
Martin Weiner

Acknowledgments

ORIGINAL EDITION

Our acknowledgments begin with the original Siddur Group, chaired in the mid-1980's by Rabbi H. Leonard Poller (*z"l*), which included some of the great scholars and liturgists of the day. They provided essential background studies for our work, based on Rabbi Joseph Glaser's (*z"l*) original vision of what a new prayerbook could be. The late poet T. Carmi combed post-biblical Hebrew literature for texts appropriate for liturgical usage. Mr. Daniel Schechter, chair of the CCAR–UAHC Joint Commission on Religious Living and a member of the Siddur Group, arranged for substantial grants from the Lilly Endowment and the Cummings Foundation to fund a three-year research study on "Lay Involvement in Liturgical Change," which yielded important guidance as we defined our criteria. By 2000 the editorial phase of this project began. In 2002, the Research Network of Tallahassee, Florida, led by Dr. Marc Gertz, designed and supervised an extensive piloting program that ultimately involved over 300 congregations and numerous conferences, conventions and focus groups.

As editor, Rabbi Elyse Frishman envisioned the overall prayerbook, contributed many selections and resource notes, and shepherded the project to its ultimate conclusion. She thanks the series of committees who worked in a truly collaborative manner to make this a genuine voice of our time. Rabbi Judith Z. Abrams, consulting editor, wrote numerous insightful and scholarly monographs on liturgical themes. Overseeing the whole, Rabbi Peter Knobel chaired the editorial committees and coordinated the enormous complexity of a prayerbook that reflects the voices of an entire movement.

A Task Force on Music created the list of musical texts. Our thanks to Cantors Richard Cohn, Alane Katzew, Marshall Portnoy and Josee Wolff; to Rabbis Lester Bronstein and Daniel Freelander, and to consultants Mark Dunn and Joyce Rosenzweig for their contributions, and to Cantors Jeffrey Klepper and Benjie Ellen Schiller who completed this phase of our work. Thanks are also due to Transcontinental Music Publications, and Joel Eglash in particular, for providing most of the musical texts in this volume. We express particular thanks to the leadership and Board of the American Conference of Cantors for enabling cantors to participate with rabbis in the manuscript review process, and for important feedback to the editorial committee.

The Siddur Publishing Team worked prodigiously to respond to the many comments submitted during the review process. In particular, abiding thanks are due to Rabbis Peter Knobel, Lawrence Hoffman, Bernard Mehlman and Elaine Zecher for their extensive contributions.

The Conference is grateful to Mr. Robert D. Rapaport for his generosity in funding the project, and for his commitment to the Jewish People. The CCAR also thanks Rabbi Howard Shapiro for bringing the important work of publishing *Mishkan T'filah* to the attention of Mr. Rapaport.

Editorial work went well beyond the appointed committees. Important manuscript contributions were made by Rabbis Howard Apothaker, A. Stanley Dreyfus, Lawrence Englander, Yoel Kahn, Lewis Kamrass, Charles Kroloff, Barton Lee, Sheldon Marder, Bernard Mehlman, James Rosenberg, Martin S. Weiner, and by Dr. Ellen Frankel, Sylvia Fuks Fried, Dara Frimmer, Howard Goldsmith, Nicole Leiser, Dr. Peggy Morrison, Adam Sol, Ellen Steinbaum and Dr. Ellen Umansky. Hundreds of additional colleagues in the rabbinate and cantorate and numerous lay persons made contributions and suggestions. Rabbis Steven Bob, Lawrence Englander, Yoel Kahn, Lewis Kamrass, Karyn Kedar, Richard Levy, Richard Sarason, Kinneret Shiryon and Martin S. Weiner provided substantial input into particular services. Rabbi Richard Sarason contributed the extensive commentary to Shabbat II services. Our work benefited particularly from Joel M. Hoffman's translations in *My People's Prayerbook* and his proofing of the Hebrew text. Dr. Yehiel Hayon also edited Hebrew texts. Rabbis Stephen Franklin, Connie Golden, and Rifat Sonsino read proof for English texts and transliterations. We are grateful to Rabbi Bernard Zlotowitz for checking source citations. We also thank Mr. Bruce Black for his many suggestions of style. The project's course was aptly guided by four CCAR Presidents, Rabbis Charles Kroloff, Martin S. Weiner, Janet Marder and Harry Danziger.

In addition, a special task force with Rabbis Lance Sussman, Elaine Zecher, Peter Knobel and Bill Rothschild oversaw the final production of the Siddur, with the help of Andrew Farber, Victor Ney, and George Davidson.

Mishkan T'filah was guided through its final stages by Rabbis Arnold Sher (Interim Executive Vice President, 2005, and Director of Placement, 1989–present) and Steven A. Fox (Executive Vice President, 2006–present), who also encouraged adoption of *Mishkan T'filah* as the unifying prayerbook of the Reform Movement. The entire CCAR staff contributed significantly to this Siddur. Rabbis Paul J. Menitoff (Executive Vice President, 1994–2005) and Elliot L. Stevens (Associate Executive Vice President, 1975– 2007) were involved through the editorial process. Deborah Smilow, CCAR's Managing Editor, devoted many hundreds of hours and meticulous care to the manuscript. In addition, Barry Nostradamus Sher, typographer, patiently went well beyond his commission to see the prayerbook through.

FROM THE EDITOR:

It is a privilege to serve, and moreover, to have one's service accepted. I am humbly grateful to the Conference and the Movement for the sacred opportunity to participate in the creation of *Mishkan T'filah.* I would like to acknowledge with deep appreciation the two congregations who helped shape my understanding of worship, The Reform Temple of Suffern – Shir Shalom (NY), and Congregation B'nai Jeshurun – The Barnert Temple (Franklin Lakes, NJ); the latter especially is thanked for truly generous support throughout the project's length. Dr. Lawrence A. Hoffman has been a mentor of liturgy and worship. Profound gratitude goes to my parents, Joan and Dr. Jack Frishman, and parents-in-law, Aviva and I. Robert Freelander, *z"l,* whose pride and love continue to inspire; to my children, Adam, Jonah and Devra, who bore the demands of the project with patience and appreciation, and whose own talents and characters fill their parents with pride; and to my husband, Rabbi Daniel H. Freelander, whose keen understanding, wisdom and warm love match the power of the sun's rising each day.

MISHKAN T'FILAH: THE JOURNAL EDITION

The idea to create *Mishkan T'filah: The Journal Edition* came directly from a suggestion made by then–NFTY Religious and Cultural Vice President Elaina Marshalek. The Central Conference of American Rabbis is delighted to bring Elaina's idea to fruition. Along the way, the CCAR received significant help and encouragement in developing this project from NFTY. We would like to especially thank Rabbi Michael Mellen, Lisa Beiber David, Paul Reichenbach, and Melissa Frey for their help and enthusiastic support, as well as the brave group of RCVPs who agreed to be part of a focus group at NFTY Convention 2009. We were privileged to be able to pilot this material at Kutz Camp during the summer of 2010, thanks to the efforts of Melissa Frey, Yoni Regev, and Rabbi Scott Weiner, as well as all the participants.

Mishkan T'filah: The Journal Edition was created by Rabbi Joel N. Abraham and Michelle Shapiro Abraham, RJE, with assistance from CCAR Press Publisher and Director, Rabbi Hara E. Person. They were guided by the ideas and feedback from the *Mishkan T'filah: The Journal Edition* Task Force, which consisted of educator colleagues from NATE, Lisa Barzilai, RJE, Stephanie Fink, RJE, and Avram Mandell, as well as Rabbis Jennifer Clayman, Richard Kellner, Paul Kipnes, Rob Nosanchuk, and Stacy Rigler, RJE, and rabbinic student PJ Schwartz. Tremendous thanks are due to all who generously helped shape this project.

This version of *Mishkan T'filah* could never have come into being without the support of Rabbi Steven A. Fox, Chief Executive of the CCAR; Rabbi Ellen Weinberg Dreyfus, President of the CCAR; Rabbi Elaine Zecher, Chair of Worship and Practice; and Rabbi Lance Sussman, Chair of the CCAR Press Committee. Thanks are also due to all those who helped move the book into its final form, including Deborah Smilow and Debra Hirsch Corman. The CCAR remains grateful to Bob and Cobey Rapaport for the support of the original edition of *Mishkan T'filah,* without which this Journal Edition would not have been possible.

A Note on Style and Usage

Mishkan T'filah offers many opportunities for diverse usage and worship styles. In the original edition of *Mishkan T'filah*, most of the prayers in this *siddur* are set as a two-page spread, with the *keva* (primary, traditional) liturgy on the right-hand page and alternative prayer choices on the left-hand side. The right-hand Hebrew text is accompanied by a faithful translation, and transliteration; the left-hand page contains poetry, prayers and *kavanot* (meditations) thematically tied to the *keva* text but reflecting diverse theological points of view. In this *Journal Edition*, the left-hand page is offered to you, for your own voice.

Layouts of the prayers invite different usage. Some passages are set to facilitate responsive readings, though all passages allow for unison offering. It is suggested that the community vary its vocal expression in prayer, with quieter or more exuberant tones at appropriate moments, as guided by the worship leader.

"Stage directions" and choreography such as when to stand or sit are minimized for maximum flexibility and with respect for varying congregational customs.

Spiritual commentaries, notes on rabbinic practices, and source citations span the bottom of the right-hand and some left-hand pages, below the liturgical frame. The heading "For those who choose" refers to Jewish customs established by long usage. While not binding, many will find them meaningful, and their use can enhance one's *kavanah*. Text sources are abbreviated: *M. Shabbat 2:1* is read as *Mishnah Shabbat, chapter 2, paragraph 1. Shabbat 2b–3a* refers to the *Talmud tractate Shabbat, page 2b to page 3a.*

Rubric headings are on the outside margins, indicating the current prayer's place within the service.

Source citations for prayers and readings by contemporary authors are found in the back of the book. Citations for the commentaries are found within the text.

A selection of musical texts is found at the back of the book. There are also blank pages in the song section for you to add in your own songs.

Doubled blank pages are found in a few select places throughout the book. These are provided in order to give you the opportunity to repond in your own voice even within certain rubrics that contain both right-hand and left-hand material.

תְּפִלּוֹת לְחוֹל

T'FILOT L'CHOL

Prayers for Weekdays

AR'VIT L'CHOL — WEEKDAY EVENING

WE ARE CALLED unto life, destiny uncertain.
Yet we offer thanks for what we know,
for health and healing, for labor and repose,
for renewal of beauty in earth and sky,
for that blend of human-holy which inspires compassion,
and for hope: eternal, promising light.

For life, for health, for hope,
for beautiful, bountiful blessing,
all praise to the Source of Being.

בָּרוּךְ אַתָּה, יְיָ,
מְקוֹר נֶפֶשׁ כָּל חַי.

Welcome

שְׁמַע וּבִרְכוֹתֶיהָ

SH'MA UVIRCHOTEHA — SH'MA AND ITS BLESSINGS

בָּרְכוּ אֶת יְיָ הַמְבֹרָךְ!
בָּרוּךְ יְיָ הַמְבֹרָךְ
לְעוֹלָם וָעֶד!

PRAISE ADONAI to whom praise is due forever!
Praised be Adonai to whom praise is due,
now and forever!

All holy acts require summoning.

For those who choose: The prayer leader at the word בָּרְכוּ *Bar'chu* (the call to worship) bends the knees and bows from the waist, and at יְיָ *Adonai* stands straight. בָּרוּךְ יְיָ *Baruch Adonai* is the communal response, whereupon the community repeats the choreography of the first line.

Bar'chu
Maariv Aravim
Ahavat Olam
Sh'ma
V'ahavta
Emet ve-Emunah
Mi Chamochah
Hashkiveinu
Chatzi Kaddish

I Imagine that you are standing in the court of the Temple in Jerusalem, and you hear the Levites chant the call to worship, "Praise Adonai to whom praise is due forever!" You answer, "Praised be Adonai, to whom praise is due, now and forever!" How do you get ready to pray? What activities, sounds, or images help you prepare for prayer?

בָּרוּךְ אַתָּה, יְיָ
אֱלֹהֵינוּ, מֶלֶךְ הָעוֹלָם,
אֲשֶׁר בִּדְבָרוֹ מַעֲרִיב עֲרָבִים,
בְּחָכְמָה פּוֹתֵחַ שְׁעָרִים,
וּבִתְבוּנָה מְשַׁנֶּה עִתִּים
וּמַחֲלִיף אֶת הַזְּמַנִּים,
וּמְסַדֵּר אֶת הַכּוֹכָבִים
בְּמִשְׁמְרוֹתֵיהֶם בָּרָקִיעַ כִּרְצוֹנוֹ.
בּוֹרֵא יוֹם וָלָיְלָה,
גּוֹלֵל אוֹר מִפְּנֵי חֹשֶׁךְ
וְחֹשֶׁךְ מִפְּנֵי אוֹר,
וּמַעֲבִיר יוֹם וּמֵבִיא לָיְלָה,
וּמַבְדִּיל בֵּין יוֹם וּבֵין לָיְלָה,
יְיָ צְבָאוֹת שְׁמוֹ.
אֵל חַי וְקַיָּם,
תָּמִיד יִמְלֹךְ עָלֵינוּ לְעוֹלָם וָעֶד.
בָּרוּךְ אַתָּה, יְיָ, הַמַּעֲרִיב עֲרָבִים.

PRAISED are You, Adonai our God, Ruler of the universe,
who speaks the evening into being,
skillfully opens the gates,
thoughtfully alters the time and changes the seasons,
and arranges the stars in their heavenly courses according to plan.
You are Creator of day and night,
rolling light away from darkness and darkness from light,
transforming day into night and distinguishing one from the other.
Adonai Tz'vaot is Your Name.
Ever-living God, may You reign continually over us into eternity.
Praise to You, Adonai, who brings on evening.

בָּרוּךְ אַתָּה, יְיָ, הַמַּעֲרִיב עֲרָבִים.

יְיָ צְבָאוֹת *Adonai Tz'vaot:* this is one of many names that help elucidate God's attributes. God designs, creates and arranges the universe with order and purpose.

P With electric lighting, we now have more control over the cycles of light and dark. How is our relationship to creation, and to the Creator, different from that of our ancestors, now that our activity is no longer restricted by the length of day?

אַהֲבַת עוֹלָם

בֵּית יִשְׂרָאֵל עַמְּךָ אָהָבְתָּ,
תּוֹרָה וּמִצְוֹת,
חֻקִּים וּמִשְׁפָּטִים אוֹתָנוּ לִמַּדְתָּ.
עַל כֵּן, יְיָ אֱלֹהֵינוּ,
בְּשָׁכְבֵנוּ וּבְקוּמֵנוּ
נָשִׂיחַ בְּחֻקֶּיךָ,
וְנִשְׂמַח בְּדִבְרֵי תוֹרָתְךָ
וּבְמִצְוֹתֶיךָ לְעוֹלָם וָעֶד.
כִּי הֵם חַיֵּינוּ וְאֹרֶךְ יָמֵינוּ
וּבָהֶם נֶהְגֶּה יוֹמָם וָלָיְלָה.
וְאַהֲבָתְךָ
אַל תָּסִיר מִמֶּנּוּ לְעוֹלָמִים.
בָּרוּךְ אַתָּה, יְיָ,
אוֹהֵב עַמּוֹ יִשְׂרָאֵל.

EVERLASTING LOVE You offered Your people Israel
by teaching us Torah and mitzvot, laws and precepts.
Therefore, Adonai our God, when we lie down and when we rise up,
we will meditate on Your laws and Your commandments.
We will rejoice in Your Torah for ever.
Day and night we will reflect on them
for they are our life and doing them lengthens our days.
Never remove Your love from us.
Praise to You, Adonai, who loves Your people Israel.

בָּרוּךְ אַתָּה, יְיָ, אוֹהֵב עַמּוֹ יִשְׂרָאֵל.

שמע ישראל יהוה

שְׁמַ֖ע יִשְׂרָאֵ֑ל יְהוָ֥ה אֱלֹהֵ֖ינוּ יְהוָ֥ה אֶחָֽד!

Hear, O Israel, Adonai is our God, Adonai is One!

בָּרוּךְ שֵׁם כְּבוֹד מַלְכוּתוֹ לְעוֹלָם וָעֶד.

Blessed is God's glorious majesty forever and ever.

בָּרְכוּ
מַעֲרִיב עֲרָבִים
אַהֲבַת עוֹלָם
שְׁמַע
וְאָהַבְתָּ
אֱמֶת וֶאֱמוּנָה
מִי־כָמֹכָה
הַשְׁכִּיבֵנוּ
חֲצִי קַדִּישׁ

וְאָהַבְתָּ אֵת יְיָ אֱלֹהֶיךָ
בְּכָל־לְבָבְךָ וּבְכָל־נַפְשְׁךָ וּבְכָל־
מְאֹדֶךָ׃ וְהָיוּ הַדְּבָרִים הָאֵלֶּה
אֲשֶׁר אָנֹכִי מְצַוְּךָ הַיּוֹם עַל־
לְבָבֶךָ׃ וְשִׁנַּנְתָּם לְבָנֶיךָ וְדִבַּרְתָּ
בָּם בְּשִׁבְתְּךָ בְּבֵיתֶךָ וּבְלֶכְתְּךָ
בַדֶּרֶךְ וּבְשָׁכְבְּךָ וּבְקוּמֶךָ׃
וּקְשַׁרְתָּם לְאוֹת עַל־יָדֶךָ וְהָיוּ
לְטֹטָפֹת בֵּין עֵינֶיךָ׃ וּכְתַבְתָּם
עַל־מְזֻזוֹת בֵּיתֶךָ וּבִשְׁעָרֶיךָ׃

לְמַעַן תִּזְכְּרוּ וַעֲשִׂיתֶם אֶת־
כָּל־מִצְוֹתָי וִהְיִיתֶם קְדֹשִׁים
לֵאלֹהֵיכֶם׃ אֲנִי יְיָ אֱלֹהֵיכֶם
אֲשֶׁר הוֹצֵאתִי אֶתְכֶם מֵאֶרֶץ
מִצְרַיִם לִהְיוֹת לָכֶם לֵאלֹהִים
אֲנִי יְיָ אֱלֹהֵיכֶם׃

YOU SHALL LOVE Adonai your God with all your heart,
with all your soul, and with all your might.
Take to heart these instructions with which I charge you this day.
Impress them upon your children.
Recite them when you stay at home and when you are away,
when you lie down and when you get up.
Bind them as a sign on your hand and let them serve as a symbol on your forehead;
inscribe them on the doorposts of your house and on your gates.

Thus you shall remember to observe all My commandments
and to be holy to your God.
I am Adonai, your God, who brought you out of the land of Egypt to be your God:
I am Adonai your God.

יְיָ אֱלֹהֵיכֶם אֱמֶת.

For those who choose: At the end of the שְׁמַע *Sh'ma*, after the words יְיָ אֱלֹהֵיכֶם *Adonai Eloheichem,* the word אֱמֶת *emet* ("true") is added as an immediate affirmation of its truth.

וְאָהַבְתָּ *V'ahavta . . . You shall love . . .* Deuteronomy 6:5–9

לְמַעַן תִּזְכְּרוּ *L'maan tizk'ru . . . Thus you shall remember . . .* Numbers 15:40–41

Bar'chu
Maariv Aravim
Ahavat Olam
Sh'ma
V'ahavta
Emet ve-Emunah
Mi Chamochah
Hashkiveinu
Chatzi Kaddish

E At some point, there may have been a member of your family you had to grow to love — a new sibling, a little-known aunt, a distant cousin. Did the expectation of love make that connection easier or harder? Stronger or weaker? Who do we love more: those we choose to love, or those we are required to love?

we love the people who are kind to us and we love our family. we love the people who we choose to love more than the people that are required to love.

בָּרְכוּ
מַעֲרִיב עֲרָבִים
אַהֲבַת עוֹלָם
שְׁמַע
וְאָהַבְתָּ
אֱמֶת וֶאֱמוּנָה
מִי־כָמֹכָה
הַשְׁכִּיבֵנוּ
חֲצִי קַדִּישׁ

אֱמֶת וֶאֱמוּנָה כָּל זֹאת
וְקַיָּם עָלֵינוּ, כִּי הוּא יְיָ אֱלֹהֵינוּ
וְאֵין זוּלָתוֹ, וַאֲנַחְנוּ יִשְׂרָאֵל עַמּוֹ.
הַפּוֹדֵנוּ מִיַּד מְלָכִים, מַלְכֵּנוּ
הַגּוֹאֲלֵנוּ מִכַּף כָּל הֶעָרִיצִים,
הָעֹשֶׂה גְדֹלוֹת עַד אֵין חֵקֶר
וְנִפְלָאוֹת עַד אֵין מִסְפָּר, הַשָּׂם
נַפְשֵׁנוּ בַּחַיִּים וְלֹא נָתַן לַמּוֹט
רַגְלֵנוּ הָעוֹשֶׂה לָנוּ נִסִּים
בְּפַרְעֹה, אוֹתוֹת וּמוֹפְתִים
בְּאַדְמַת בְּנֵי חָם. וַיּוֹצֵא אֶת
עַמּוֹ יִשְׂרָאֵל מִתּוֹכָם לְחֵרוּת
עוֹלָם. וְרָאוּ בָנָיו גְּבוּרָתוֹ, שִׁבְּחוּ
וְהוֹדוּ לִשְׁמוֹ. וּמַלְכוּתוֹ בְּרָצוֹן
קִבְּלוּ עֲלֵיהֶם. מֹשֶׁה וּמִרְיָם
וּבְנֵי יִשְׂרָאֵל לְךָ עָנוּ שִׁירָה
בְּשִׂמְחָה רַבָּה, וְאָמְרוּ כֻלָּם:

ALL THIS WE HOLD to be true and trustworthy for us.
You alone are our God, and we are Israel Your people.
You are our Sovereign and Savior,
who delivers us from oppressors' hands
and saves us from tyrants' fists.
You work wonders without number, marvels beyond count.
You give us life and steady our footsteps.
You performed miracles for us before Pharaoh,
signs and wonders in the land of the Egyptians;
You led Your people Israel out from their midst to freedom for all time.
When Your children witnessed Your dominance
they praised Your Name in gratitude.
And they accepted Your sovereignty —
Moses, Miriam and all Israel sang to You together,
lifting their voices joyously:

הָעֹשֶׂה גְדֹלוֹת *Haoseh g'dolot . . . You work wonders . . .* Job 9:10

הַשָּׂם נַפְשֵׁנוּ בַּחַיִּים *Hasam nafsheinu bachayim . . . You give us life . . .* Psalm 66:9

מִי־כָמְֽכָה בָּאֵלִם, יְיָ!
מִי כָּמְֽכָה נֶאְדָּר בַּקֹּֽדֶשׁ,
נוֹרָא תְהִלֹּת, עֹֽשֵׂה פֶֽלֶא!

מַלְכוּתְךָ רָאוּ בָנֶֽיךָ,
בּוֹקֵֽעַ יָם לִפְנֵי מֹשֶׁה וּמִרְיָם.
זֶה אֵלִי, עָנוּ וְאָמְֽרוּ,
יְיָ יִמְלֹךְ לְעֹלָם וָעֶד!

וְנֶאֱמַר: כִּי פָדָה יְיָ אֶת יַעֲקֹב
וּגְאָלוֹ מִיַּד חָזָק מִמֶּֽנּוּ.
בָּרוּךְ אַתָּה, יְיָ, גָּאַל יִשְׂרָאֵל.

WHO IS LIKE YOU, O God,
among the gods that are worshipped?
Who is like You, majestic in holiness,
awesome in splendor, working wonders?

Your children witnessed Your sovereignty,
the sea splitting before Moses and Miriam.
"This is our God!" they cried.
"Adonai will reign forever and ever!"

Thus it is said,
"Adonai redeemed Jacob,
from a hand stronger
than his own."
Praised are You, Adonai, for redeeming Israel.

בָּרוּךְ אַתָּה, יְיָ, גָּאַל יִשְׂרָאֵל.

מִי־כָמֹֽכָה *Mi chamochah . . . Who is like you . . .* Exodus 15:11

זֶה אֵלִי *Zeh Eli . . . This is our God . . .* Exodus 15:2

יְיָ יִמְלֹךְ *Adonai yimloch . . . Adonai will reign . . .* Exodus 15:18

כִּי פָדָה יְיָ *Ki fadah Adonai . . . Adonai redeemed . . .* Jeremiah 31:10

Bar'chu
Maariv Aravim
Ahavat Olam
Sh'ma
V'ahavta
Emet ve-Emunah
Mi Chamochah
Hashkiveinu
Chatzi Kaddish

בָּרְכוּ

מַעֲרִיב עֲרָבִים

אַהֲבַת עוֹלָם

שְׁמַע

וְאָהַבְתָּ

אֱמֶת וֶאֱמוּנָה

מִי־כָמֹכָה

הַשְׁכִּיבֵנוּ

חֲצִי קַדִּישׁ

הַשְׁכִּיבֵנוּ, יְיָ אֱלֹהֵינוּ,

לְשָׁלוֹם, וְהַעֲמִידֵנוּ שׁוֹמְרֵנוּ לְחַיִּים,

וּפְרֹשׂ עָלֵינוּ סֻכַּת שְׁלוֹמֶךָ,

וְתַקְּנֵנוּ בְּעֵצָה טוֹבָה מִלְּפָנֶיךָ,

וְהוֹשִׁיעֵנוּ לְמַעַן שְׁמֶךָ.

וְהָגֵן בַּעֲדֵנוּ,

וְהָסֵר מֵעָלֵינוּ אוֹיֵב, דֶּבֶר,

וְחֶרֶב, וְרָעָב, וְיָגוֹן,

וְהַרְחֵק מִמֶּנּוּ עָוֹן וָפֶשַׁע.

וּבְצֵל כְּנָפֶיךָ תַּסְתִּירֵנוּ,

כִּי אֵל שׁוֹמְרֵנוּ וּמַצִּילֵנוּ אָתָּה,

כִּי אֵל חַנּוּן וְרַחוּם אָתָּה.

וּשְׁמֹר צֵאתֵנוּ וּבוֹאֵנוּ

לְחַיִּים וּלְשָׁלוֹם

מֵעַתָּה וְעַד עוֹלָם.

בָּרוּךְ אַתָּה, יְיָ,

שׁוֹמֵר עַמּוֹ יִשְׂרָאֵל לָעַד.

GRANT, O GOD, that we lie down in peace,
and raise us up, our Guardian, to life renewed.
Spread over us the shelter of Your peace.
Guide us with Your good counsel;
for Your Name's sake, be our help.
Shield and shelter us beneath the shadow of Your wings.
Defend us against enemies, illness, war, famine and sorrow.
Distance us from wrongdoing.
For You, God, watch over us and deliver us.
For You, God, are gracious and merciful.
Guard our going and coming, to life and to peace evermore.

Blessed are You, Adonai, Guardian of Israel.

בָּרוּךְ אַתָּה, יְיָ, שׁוֹמֵר עַמּוֹ יִשְׂרָאֵל לָעַד.

Grant, O God, that we lie down in peace . . . Following a reading from *Seder Rav Amram*, our first known comprehensive prayerbook, circa 860 C.E.

בָּרְכוּ
מַעֲרִיב עֲרָבִים
אַהֲבַת עוֹלָם
שְׁמַע
וְאָהַבְתָּ
אֱמֶת וֶאֱמוּנָה
מִי־כָמֹכָה
הַשְׁכִּיבֵנוּ
חֲצִי קַדִּישׁ

יִתְגַּדַּל וְיִתְקַדַּשׁ שְׁמֵהּ רַבָּא
בְּעָלְמָא דִּי בְרָא כִרְעוּתֵהּ,
וְיַמְלִיךְ מַלְכוּתֵהּ
בְּחַיֵּיכוֹן וּבְיוֹמֵיכוֹן
וּבְחַיֵּי דְכָל בֵּית יִשְׂרָאֵל,
בַּעֲגָלָא וּבִזְמַן קָרִיב,
וְאִמְרוּ: אָמֵן.

יְהֵא שְׁמֵהּ רַבָּא מְבָרַךְ
לְעָלַם וּלְעָלְמֵי עָלְמַיָּא.

יִתְבָּרַךְ וְיִשְׁתַּבַּח וְיִתְפָּאַר
וְיִתְרוֹמַם וְיִתְנַשֵּׂא,
וְיִתְהַדַּר וְיִתְעַלֶּה וְיִתְהַלַּל
שְׁמֵהּ דְּקֻדְשָׁא בְּרִיךְ הוּא,
לְעֵלָּא מִן כָּל בִּרְכָתָא וְשִׁירָתָא,
תֻּשְׁבְּחָתָא וְנֶחֱמָתָא,
דַּאֲמִירָן בְּעָלְמָא, וְאִמְרוּ: אָמֵן.

EXALTED and hallowed be God's great name,
in the world which God created, according to plan.
May God's majesty be revealed in the days of our lifetime
and the life of all Israel —
speedily, imminently.
To which we say: Amen.

Blessed be God's great name to all eternity.

Blessed, praised, honored, exalted,
extolled, glorified, adored, and lauded
be the name of the Holy Blessed One,
beyond all earthly words and songs of blessing, praise, and comfort.
To which we say: Amen.

For Weekday T'filah, turn to pages 74–75.

The קַדִּישׁ *Kaddish* is marked by long strings of synonyms of praise. The rhythmic repetition of these words is meant to aid one in achieving a higher meditational state. *Judith Z. Abrams*

Bar'chu
Maariv Aravim
Ahavat Olam
Sh'ma
V'ahavta
Emet ve-Emunah
Mi Chamochah
Hashkiveinu
Chatzi Kaddish

P The *Chatzi Kaddish* refers to God using the words "Blessed, praised, honored, exalted, extolled, glorified, adored, and lauded." Is this too much, even for God? What do you think this prayer is trying to say about the limits of language and the boundlessness of the Divine?

SHACHARIT L'CHOL — WEEKDAY MORNING

IN THE MORNING, before this day's journey begins,
I offer thanks before You, God,
that just as You found me worthy
to gaze upon the sun in the east,
so I will merit seeing it in the west.

And when darkness descends,
may it be Your will to grace me
with another dawning of light.

WITHIN THESE WALLS we sit surrounded by numberless generations.
Our ancestors built the synagogue as a visible sign of God's presence in their midst.
Throughout our long history, our endless wanderings it has endured,
a beacon of truth, love, and justice for all humanity.
Its presence guided our ancestors to lives of righteousness,
holding up to them a vision of their truest selves.
Now we, in our turn, come into this sanctuary to affirm the sacredness of our lives.

May holiness wrap around us as we cross its threshold.
May we enter this place in peace.

בִּרְכוֹת הַשַּׁחַר

BIRCHOT HASHACHAR — MORNING BLESSINGS

מוֹדֶה / מוֹדָה אֲנִי לְפָנֶיךָ,
מֶלֶךְ חַי וְקַיָּם,
שֶׁהֶחֱזַרְתָּ בִּי נִשְׁמָתִי בְּחֶמְלָה,
רַבָּה אֱמוּנָתֶךָ.

I OFFER THANKS to You,
ever-living Sovereign,
that You have restored my soul to me in mercy:
How great is Your trust.

Modeh / Modah Ani
Tzitzit
T'fillin
Mah Tovu
Asher Yatzar
Elohai N'shamah
Nisim B'chol Yom
Laasok
V'haarev Na
Eilu D'varim
Kaddish D'Rabanan

I Imagine life without a soul, without that spark or animation. Then feel that spark return and ignite a flame. Let that feeling form a prayer.

FOR THOSE WHO WEAR TALLIT

AS I WRAP myself in the tallit,
I fulfill the mitzvah of my Creator.

Before putting on tallit

בָּרְכִי נַפְשִׁי אֶת יְיָ!
יְיָ אֱלֹהַי, גָּדַֽלְתָּ מְּאֹד,
הוֹד וְהָדָר לָבָֽשְׁתָּ.
עֹֽטֶה אוֹר כַּשַּׂלְמָה,
נוֹטֶה שָׁמַֽיִם כַּיְרִיעָה.

BLESS, ADONAI, O my soul!
Adonai my God, how great You are.
You are robed in glory and majesty,
wrapping Yourself in light as in a garment,
spreading forth the heavens like a curtain.

בָּרוּךְ אַתָּה, יְיָ
אֱלֹהֵֽינוּ, מֶֽלֶךְ הָעוֹלָם,
אֲשֶׁר קִדְּשָֽׁנוּ בְּמִצְוֹתָיו
וְצִוָּֽנוּ לְהִתְעַטֵּף בַּצִּיצִת.

BLESSED ARE YOU, Adonai our God,
Sovereign of the universe,
who hallows us with mitzvot,
commanding us to wrap ourselves in the fringes.

וְצִוָּֽנוּ לְהִתְעַטֵּף בַּצִּיצִת *v'tzivanu l'hitateif batzitzit . . . commanding us to wrap ourselves in the fringes.* This mitzvah is drawn from Numbers 15:38–39.

Modeh / Modah Ani

Tzitzit

T'fillin

Mah Tovu

Asher Yatzar

Elohai N'shamah

Nisim B'chol Yom

Laasok

V'haarev Na

Eilu D'varim

Kaddish D'Rabanan

מוֹדֶה / מוֹדָה אֲנִי

צִיצִת

תְּפִלִּין

מַה־טֹּבוּ

אֲשֶׁר יָצַר

אֱלֹהַי נְשָׁמָה

נִסִּים בְּכָל יוֹם

לַעֲסוֹק

וְהַעֲרֶב־נָא

אֵלּוּ דְבָרִים

קַדִּישׁ דְּרַבָּנָן

FOR THOSE WHO WEAR T'FILLIN

When placing on the arm

בָּרוּךְ אַתָּה, יְיָ

אֱלֹהֵינוּ, מֶלֶךְ הָעוֹלָם,

אֲשֶׁר קִדְּשָׁנוּ בְּמִצְוֹתָיו

וְצִוָּנוּ לְהָנִיחַ תְּפִלִּין.

BLESSED ARE YOU, Adonai our God, Sovereign of the universe.
who hallows us with mitzvot,
commanding us to wind the t'fillin.

After placing on the forehead

בָּרוּךְ אַתָּה, יְיָ

אֱלֹהֵינוּ, מֶלֶךְ הָעוֹלָם,

אֲשֶׁר קִדְּשָׁנוּ בְּמִצְוֹתָיו

וְצִוָּנוּ עַל מִצְוַת תְּפִלִּין.

בָּרוּךְ שֵׁם כְּבוֹד מַלְכוּתוֹ

לְעוֹלָם וָעֶד.

BLESSED ARE YOU, Adonai our God, Sovereign of the universe.
You have sanctified us through Your commandments,
commanding us concerning t'fillin.
Blessed is God's glorious majesty forever and ever.

When winding three times around the middle finger

וְאֵרַשְׂתִּיךְ לִי לְעוֹלָם,

וְאֵרַשְׂתִּיךְ לִי בְּצֶדֶק וּבְמִשְׁפָּט

וּבְחֶסֶד וּבְרַחֲמִים.

וְאֵרַשְׂתִּיךְ לִי בֶּאֱמוּנָה,

וְיָדַעַתְּ אֶת יְיָ.

I WILL BETROTH you to Me forever;
I will betroth you to Me
in righteousness and in justice, in kindness and in mercy.
I will betroth you to Myself in faithfulness,
and you shall know Adonai.

בָּרוּךְ שֵׁם כְּבוֹד *Baruch shem k'vod . . . M. Yoma 3:8; 8:2*

וְאֵרַשְׂתִּיךְ לִי לְעוֹלָם *V'eirastich li l'olam . . . I will betroth you to Me forever . . .* Hosea 2:21–22

Modeh / Modah Ani
Tzitzit
T'fillin
Mah Tovu
Asher Yatzar
Elohai N'shamah
Nisim B'chol Yom
Laasok
V'haarev Na
Eilu D'varim
Kaddish D'Rabanan

T'fillin illustrate Judaism's effort to synthesize faith and deed. Wrapping words of Torah around one's body is a vivid reminder not merely to recite or to study Torah, but to follow its instructions.

Wrapping sacred words around one's head and arm is a reminder that the body, too, is sacred.

The *t'fillin* contain passages from Deuteronomy and Numbers known as "the *Sh'ma* and three paragraphs." Since the first word after the *Sh'ma* begins "*V'ahavta,* You shall love," *t'fillin* suggest a relationship of affection and intimacy with God. This is best indicated by the way in which the straps are wound around the fingers and by the verses recited at that point. The straps wound around the fingers remind one of a wedding band, and the seven windings around the arm have been taken as a symbol of the seven wedding blessings. During the seventeenth century, it became customary to recite the magnificent wedding formula of God and Israel found in the Book of Hosea, 2:21-22. *adapted from Reuven Hammer*

מַה־טֹּבוּ אֹהָלֶיךָ, יַעֲקֹב,
מִשְׁכְּנֹתֶיךָ, יִשְׂרָאֵל!

וַאֲנִי בְּרֹב חַסְדְּךָ
אָבוֹא בֵיתֶךָ,
אֶשְׁתַּחֲוֶה אֶל־הֵיכַל קָדְשְׁךָ
בְּיִרְאָתֶךָ.

יְיָ, אָהַבְתִּי מְעוֹן בֵּיתֶךָ
וּמְקוֹם מִשְׁכַּן כְּבוֹדֶךָ.

וַאֲנִי אֶשְׁתַּחֲוֶה וְאֶכְרָעָה,
אֶבְרְכָה לִפְנֵי־יְיָ עֹשִׂי.

וַאֲנִי תְפִלָּתִי־לְךָ, יְיָ,
עֵת רָצוֹן.
אֱלֹהִים, בְּרָב־חַסְדֶּךָ,
עֲנֵנִי בֶּאֱמֶת יִשְׁעֶךָ.

HOW FAIR are your tents, O Jacob,
your dwellings, O Israel.

I, through Your abundant love, enter Your house;
I bow down in awe at Your holy temple.

Adonai, I love Your temple abode,
the dwelling-place of Your glory.

I will humbly bow down low before Adonai, my Maker.

As for me, may my prayer come to You, Adonai, at a favorable time.
O God, in Your abundant faithfulness, answer me with Your sure deliverance.

The opening words of this passage are from Numbers 24:5 where they are recited by Balaam, the foreign prophet who was commissioned to curse the children of Israel. When he opened his mouth, blessings emerged instead of curses.

מַה־טֹּבוּ *Mah tovu . . . How fair . . .* Numbers 24:5

וַאֲנִי בְּרֹב חַסְדְּךָ *Vaani b'rov chasd'cha . . . I, through Your abundant love . . .* Psalm 5:8

יְיָ, אָהַבְתִּי *Adonai, ahavti . . . Adonai, I love . . .* Psalm 26:8

וַאֲנִי תְפִלָּתִי *Vaani t'filati . . . As for me, may my prayer . . .* Psalm 69:14

I Stand with Balaam as he overlooks the tents of the Israelite people, a curse on your lips. What do you see in this ragtag gathering of ex-slaves that causes the words to die on your lips and make you instead utter the blessing: "How fair are your tents, O Jacob"?

בָּרוּךְ אַתָּה, יְיָ
אֱלֹהֵינוּ, מֶלֶךְ הָעוֹלָם,
אֲשֶׁר יָצַר אֶת הָאָדָם בְּחָכְמָה
וּבָרָא בוֹ נְקָבִים, נְקָבִים,
חֲלוּלִים, חֲלוּלִים.
גָּלוּי וְיָדְוּעַ לִפְנֵי כִסֵּא כְבוֹדֶךָ
שֶׁאִם יִפָּתֵחַ אֶחָד מֵהֶם,
אוֹ יִסָּתֵם אֶחָד מֵהֶם,
אִי אֶפְשַׁר לְהִתְקַיֵּם
וְלַעֲמוֹד לְפָנֶיךָ.
בָּרוּךְ אַתָּה, יְיָ,
רוֹפֵא כָל בָּשָׂר וּמַפְלִיא לַעֲשׂוֹת.

PRAISE TO YOU, Adonai,
our God, Sovereign of the universe,
who formed the human body with skill
creating the body's many pathways and openings.
It is well known before Your throne of glory
that if one of them be wrongly opened or closed,
it would be impossible to endure and stand before You.
Blessed are You, Adonai, who heals all flesh, working wondrously.

בָּרוּךְ אַתָּה, יְיָ, רוֹפֵא כָל בָּשָׂר וּמַפְלִיא לַעֲשׂוֹת.

P If we are each created *b'tzelem Elohim* — in the image of God — then imagine how complex and integrated God must be. What are the metaphoric arteries, organs, and networks that make up the Divine?

אֱלֹהַי, נְשָׁמָה שֶׁנָּתַתָּ בִּי
טְהוֹרָה הִיא.
אַתָּה בְרָאתָהּ, אַתָּה יְצַרְתָּהּ,
אַתָּה נְפַחְתָּהּ בִּי,
וְאַתָּה מְשַׁמְּרָהּ בְּקִרְבִּי.
כָּל זְמַן שֶׁהַנְּשָׁמָה בְּקִרְבִּי,
מוֹדֶה / מוֹדָה אֲנִי לְפָנֶיךָ,
יְיָ אֱלֹהַי
וֵאלֹהֵי אֲבוֹתַי וְאִמּוֹתַי,
רִבּוֹן כָּל הַמַּעֲשִׂים,
אֲדוֹן כָּל הַנְּשָׁמוֹת.
בָּרוּךְ אַתָּה, יְיָ,
אֲשֶׁר בְּיָדוֹ נֶפֶשׁ כָּל חָי
וְרוּחַ כָּל בְּשַׂר אִישׁ.

MY GOD, the soul You have given me is pure.
You created it, You shaped it, You breathed it into me
and You protect it within me.
For as long as my soul is within me,
I offer thanks to You,
Adonai, my God
and God of my ancestors
Source of all Creation, Sovereign of all souls.
Praised are You, Adonai,
in whose hand is every living soul and the breath of humankind.

בָּרוּךְ אַתָּה, יְיָ, אֲשֶׁר בְּיָדוֹ נֶפֶשׁ כָּל חָי וְרוּחַ כָּל בְּשַׂר אִישׁ.

אֱלֹהַי, נְשָׁמָה *Elohai, n'shamah . . . My God, the soul . . . based on B'rachot 60b*

אֲשֶׁר בְּיָדוֹ *Asher b'yado . . . in whose hand . . .* Job 12:10

נִסִּים בְּכָל יוֹם

NISIM B'CHOL YOM — FOR DAILY MIRACLES

For awakening

בָּרוּךְ אַתָּה, יְיָ
אֱלֹהֵינוּ, מֶלֶךְ הָעוֹלָם,
אֲשֶׁר נָתַן לַשֶּׂכְוִי בִינָה
לְהַבְחִין בֵּין יוֹם וּבֵין לָיְלָה.

PRAISE TO YOU, Adonai our God, Sovereign of the universe,
who has given the mind the ability to distinguish day from night.

For vision

בָּרוּךְ אַתָּה, יְיָ
אֱלֹהֵינוּ, מֶלֶךְ הָעוֹלָם,
פּוֹקֵחַ עִוְרִים.

PRAISE TO YOU, Adonai our God, Sovereign of the universe,
who opens the eyes of the blind.

For the ability to stretch

בָּרוּךְ אַתָּה, יְיָ
אֱלֹהֵינוּ, מֶלֶךְ הָעוֹלָם,
מַתִּיר אֲסוּרִים.

PRAISE TO YOU, Adonai our God, Sovereign of the universe,
who frees the captive.

For rising to the new day

בָּרוּךְ אַתָּה, יְיָ
אֱלֹהֵינוּ, מֶלֶךְ הָעוֹלָם,
זוֹקֵף כְּפוּפִים.

PRAISE TO YOU, Adonai our God, Sovereign of the universe,
who lifts up the fallen.

נִסִּים בְּכָל יוֹם *Nisim b'chol yom . . . For daily miracles . . .* These morning blessings evoke wonder at awakening to physical life: we open our eyes, clothe our bodies, and walk again with purpose; spiritual life also, we are created in God's image, are free human beings, and as Jews, celebrate the joy and destiny of our people, Israel.

Ashkenazi tradition places the "identity" blessings near the beginning; Maimonides puts them at the end.

Though they are intended literally, we may perceive each blessing spiritually.

Inspiration for blessings three to five comes from Psalm 146:7–8.

נִסִּים בְּכָל יוֹם

NISIM B'CHOL YOM — FOR DAILY MIRACLES

For firm earth to stand upon

בָּרוּךְ אַתָּה, יְיָ
אֱלֹהֵינוּ, מֶלֶךְ הָעוֹלָם,
רוֹקַע הָאָרֶץ עַל הַמָּיִם.

PRAISE TO YOU, Adonai our God, Sovereign of the universe,
who stretches the earth over the waters.

For the gift of motion

בָּרוּךְ אַתָּה, יְיָ
אֱלֹהֵינוּ, מֶלֶךְ הָעוֹלָם,
הַמֵּכִין מִצְעֲדֵי גָבֶר.

PRAISE TO YOU, Adonai our God, Sovereign of the universe,
who strengthens our steps.

For clothing the body

בָּרוּךְ אַתָּה, יְיָ
אֱלֹהֵינוּ, מֶלֶךְ הָעוֹלָם,
מַלְבִּישׁ עֲרֻמִּים.

PRAISE TO YOU, Adonai our God, Sovereign of the universe,
who clothes the naked.

For renewed enthusiasm for life

בָּרוּךְ אַתָּה, יְיָ
אֱלֹהֵינוּ, מֶלֶךְ הָעוֹלָם,
הַנּוֹתֵן לַיָּעֵף כֹּחַ.

PRAISE TO YOU, Adonai our God, Sovereign of the universe,
who gives strength to the weary.

For reawakening

בָּרוּךְ אַתָּה, יְיָ
אֱלֹהֵינוּ, מֶלֶךְ הָעוֹלָם,
הַמַּעֲבִיר שֵׁנָה מֵעֵינַי,
וּתְנוּמָה מֵעַפְעַפָּי.

PRAISE TO YOU, Adonai our God, Sovereign of the universe,
who removes sleep from the eyes, slumber from the eyelids.

נִסִּים בְּכָל יוֹם

NISIM B'CHOL YOM — FOR DAILY MIRACLES

For being in the image of God

בָּרוּךְ אַתָּה, יְיָ
אֱלֹהֵינוּ, מֶלֶךְ הָעוֹלָם,
שֶׁעָשַׂנִי בְּצֶלֶם אֱלֹהִים.

PRAISE TO YOU, Adonai our God, Sovereign of the universe,
who made me in the image of God.

For being a free person

בָּרוּךְ אַתָּה, יְיָ
אֱלֹהֵינוּ, מֶלֶךְ הָעוֹלָם,
שֶׁעָשַׂנִי בֶּן/בַּת חוֹרִין.

PRAISE TO YOU, Adonai our God, Sovereign of the universe,
who has made me free.

For being a Jew

בָּרוּךְ אַתָּה, יְיָ
אֱלֹהֵינוּ, מֶלֶךְ הָעוֹלָם,
שֶׁעָשַׂנִי יִשְׂרָאֵל.

PRAISE TO YOU, Adonai our God, Sovereign of the universe,
who has made me a Jew.

For purpose

בָּרוּךְ אַתָּה, יְיָ
אֱלֹהֵינוּ, מֶלֶךְ הָעוֹלָם,
אוֹזֵר יִשְׂרָאֵל בִּגְבוּרָה.

PRAISE TO YOU, Adonai our God, Sovereign of the universe,
who girds Israel with strength.

For harmony

בָּרוּךְ אַתָּה, יְיָ
אֱלֹהֵינוּ, מֶלֶךְ הָעוֹלָם,
עוֹטֵר יִשְׂרָאֵל בְּתִפְאָרָה.

PRAISE TO YOU, Adonai our God, Sovereign of the universe,
who crowns Israel with splendor.

שֶׁעָשַׂנִי יִשְׂרָאֵל *She-asani Yisrael . . . For being a Jew.* Israel was the name Jacob acquired after wrestling with the angel, and this name became that of our people; we are the Children of Israel. The name Israel implies wrestling with God; to be a Jew and have faith in God is an ongoing challenge, and we are encouraged to question and delve into the nature of a faithful life.

Modeh / Modah Ani

Tzitzit

T'fillin

Mah Tovu

Asher Yatzar

Elohai N'shamah

Nisim B'chol Yom

Laasok

V'haarev Na

Eilu D'varim

Kaddish D'Rabanan

בָּרוּךְ אַתָּה, יְיָ
אֱלֹהֵינוּ, מֶלֶךְ הָעוֹלָם,
אֲשֶׁר קִדְּשָׁנוּ בְּמִצְוֹתָיו
וְצִוָּנוּ לַעֲסוֹק בְּדִבְרֵי תוֹרָה.

BLESSED ARE YOU, Adonai our God,
Sovereign of the universe,
who hallows us with mitzvot,
commanding us to engage with words of Torah.

וְהַעֲרֶב־נָא יְיָ אֱלֹהֵינוּ
אֶת־דִּבְרֵי תוֹרָתְךָ בְּפִינוּ,
וּבְפִי עַמְּךָ בֵּית יִשְׂרָאֵל,
וְנִהְיֶה אֲנַחְנוּ וְצֶאֱצָאֵינוּ,
וְצֶאֱצָאֵי עַמְּךָ בֵּית יִשְׂרָאֵל,
כֻּלָּנוּ יוֹדְעֵי שְׁמֶךָ,
וְלוֹמְדֵי תוֹרָתֶךָ לִשְׁמָהּ.
בָּרוּךְ אַתָּה, יְיָ,
הַמְלַמֵּד תּוֹרָה לְעַמּוֹ יִשְׂרָאֵל.

O ADONAI, OUR GOD,
let the words of Torah be sweet in our mouths,
and the mouths of Your people Israel,
so that we, our descendants
and the descendants of all Your people Israel may know You,
by studying Your Torah for its own sake.
Blessed are You, Adonai, who teaches Torah to Your people Israel.

בָּרוּךְ אַתָּה יְיָ, הַמְלַמֵּד תּוֹרָה לְעַמּוֹ יִשְׂרָאֵל.
Baruch atah Adonai, ham'lameid Torah l'amo Yisrael.

"Descendants" includes all women and men who embrace the Jewish people and faith.

The two blessings above (derived from *B'rachot 11b*) are both Torah blessings and are intended to introduce a moment of Torah study, which could be done at this time.

אֵלוּ דְבָרִים שֶׁאֵין לָהֶם שִׁעוּר,
שֶׁאָדָם אוֹכֵל פֵּרוֹתֵיהֶם
בָּעוֹלָם הַזֶּה
וְהַקֶּרֶן קַיֶּמֶת לוֹ לָעוֹלָם הַבָּא.
וְאֵלוּ הֵן:
כִּבּוּד אָב וָאֵם,
וּגְמִילוּת חֲסָדִים,
וְהַשְׁכָּמַת בֵּית הַמִּדְרָשׁ
שַׁחֲרִית וְעַרְבִית,
וְהַכְנָסַת אוֹרְחִים,
וּבִקוּר חוֹלִים,
וְהַכְנָסַת כַּלָּה, וּלְוָיַת הַמֵּת,
וְעִיּוּן תְּפִלָּה,
וַהֲבָאַת שָׁלוֹם
בֵּין אָדָם לַחֲבֵרוֹ,
וְתַלְמוּד תּוֹרָה כְּנֶגֶד כֻּלָּם.

מוֹדָה / מוֹדֶה אֲנִי
צִיצִת
תְּפִלִּין
מַה־טֹּבוּ
אֲשֶׁר יָצַר
אֱלֹהַי נְשָׁמָה
נִסִּים בְּכָל יוֹם
לַעֲסוֹק
וְהַעֲרֶב־נָא
אֵלוּ דְבָרִים
קַדִּישׁ דְּרַבָּנָן

THESE ARE THINGS that are limitless,
of which a person enjoys the fruit of this world,
while the principal remains in the world to come.
They are: honoring one's father and mother,
engaging in deeds of compassion,
arriving early for study, morning and evening,
dealing graciously with guests, visiting the sick,
providing for the wedding couple,
accompanying the dead for burial,
being devoted in prayer,
and making peace among people.
But the study of Torah encompasses them all.

וְהַשְׁכָּמַת בֵּית הַמִּדְרָשׁ *V'hashkamat beit hamidrash . . . arriving early for study . . .* The Rabbis understood this to convey enthusiasm and earnestness. It is not sufficient merely to attend; one's full attention is required. *Yoel Kahn*

תַּלְמוּד תּוֹרָה *Talmud Torah . . . the study of Torah* offers the knowledge of what is right and how to live justly. Jewish study includes the expectation that the lessons will be applied to life.

אֵלוּ דְבָרִים *Eilu d'varim . . . These are things . . . based on Peah 1:1*

שֶׁאָדָם אוֹכֵל *She-adam ocheil . . . of which a person enjoys . . . Shabbat 127a*

P This prayer teaches that these are deeds whose reward is without measure. However, the prayer goes on to teach that we receive benefit both here and in the world to come. Do you think that people need to receive rewards for doing the right thing? Should people be punished for doing the wrong thing? Why or why not?

מוֹדֶה / מוֹדָה אֲנִי

צִיצִת

תְּפִלִּין

מַה־טֹּבוּ

אֲשֶׁר יָצַר

אֱלֹהַי נְשָׁמָה

נִסִּים בְּכָל יוֹם

לַעֲסוֹק

וְהַעֲרֶב־נָא

אֵלּוּ דְבָרִים

קַדִּישׁ דְּרַבָּנָן

יִתְגַּדַּל וְיִתְקַדַּשׁ שְׁמֵהּ רַבָּא
בְּעָלְמָא דִּי בְרָא כִרְעוּתֵהּ,
וְיַמְלִיךְ מַלְכוּתֵהּ
בְּחַיֵּיכוֹן וּבְיוֹמֵיכוֹן
וּבְחַיֵּי דְכָל בֵּית יִשְׂרָאֵל,
בַּעֲגָלָא וּבִזְמַן קָרִיב,
וְאִמְרוּ: אָמֵן.

יְהֵא שְׁמֵהּ רַבָּא מְבָרַךְ
לְעָלַם וּלְעָלְמֵי עָלְמַיָּא.

יִתְבָּרַךְ וְיִשְׁתַּבַּח וְיִתְפָּאַר
וְיִתְרוֹמַם וְיִתְנַשֵּׂא,
וְיִתְהַדַּר וְיִתְעַלֶּה וְיִתְהַלַּל
שְׁמֵהּ דְּקֻדְשָׁא בְּרִיךְ הוּא,
לְעֵלָּא מִן כָּל בִּרְכָתָא וְשִׁירָתָא,
תֻּשְׁבְּחָתָא וְנֶחֱמָתָא,
דַּאֲמִירָן בְּעָלְמָא, וְאִמְרוּ: אָמֵן.

EXALTED and hallowed be God's great name,
in the world which God created, according to plan.
May God's majesty be revealed in the days of our lifetime
and the life of all Israel — speedily, imminently.
To which we say: Amen.

Blessed be God's great name to all eternity.

Blessed, praised, honored, exalted,
extolled, glorified, adored, and lauded
be the name of the Holy Blessed One,
beyond all earthly words and songs of blessing, praise, and comfort.
To which we say: Amen.

Kaddish D'Rabanan continues on pages 48–49.

Current scholarship sees the קַדִּישׁ דְּרַבָּנָן *Kaddish d'Rabanan* as but one of many alternative early versions of the *Kaddish.* It emerged in an oral form in the first or second century. Like other forms of the *Kaddish,* it is an elaborate praise of God, calling for the coming of God's ultimate dominion. Its unique name, קַדִּישׁ דְּרַבָּנָן *Kaddish d'Rabanan ("Kaddish of the Rabbis"),* reflects its central paragraph, acknowledging those who study Torah, indicating the role of Torah study as both an intellectual and a spiritual activity. This Kaddish concludes text study that takes place during worship and other occasions.

עַל יִשְׂרָאֵל וְעַל רַבָּנָן,
וְעַל תַּלְמִידֵיהוֹן
וְעַל כָּל תַּלְמִידֵי תַלְמִידֵיהוֹן,
וְעַל כָּל מָן דְּעָסְקִין בְּאוֹרַיְתָא,
דִּי בְּאַתְרָא הָדֵן וְדִי
בְּכָל אֲתַר וַאֲתַר,
יְהֵא לְהוֹן וּלְכוֹן שְׁלָמָא רַבָּא,
חִנָּא וְחִסְדָּא וְרַחֲמִין,
וְחַיִּין אֲרִיכִין,
וּמְזוֹנֵי רְוִיחֵי, וּפֻרְקָנָא,
מִן קֳדָם אֲבוּהוֹן
דְּבִשְׁמַיָּא וְאַרְעָא וְאִמְרוּ: אָמֵן.
יְהֵא שְׁלָמָא רַבָּא מִן שְׁמַיָּא,
וְחַיִּים טוֹבִים עָלֵינוּ וְעַל כָּל
יִשְׂרָאֵל וְאִמְרוּ: אָמֵן.
עֹשֶׂה שָׁלוֹם בִּמְרוֹמָיו
הוּא בְּרַחֲמָיו יַעֲשֶׂה שָׁלוֹם
עָלֵינוּ וְעַל כָּל יִשְׂרָאֵל, וְעַל כָּל
יוֹשְׁבֵי תֵבֵל, וְאִמְרוּ: אָמֵן.

GOD OF HEAVEN AND EARTH, grant abundant peace
to our people Israel and their rabbis, to our teachers and their disciples,
and to all who engage in the study of Torah here and everywhere.
Let there be for them and for us all, grace, love, and compassion,
a full life, ample sustenance, and salvation from God, and let us respond: Amen.

For us and all Israel, may the blessing of peace and the promise of life come true,
and let us respond: Amen.
May the One who causes peace to reign in the high heavens,
let peace descend on us, and on all Israel, and on all the world,
and let us respond: Amen.

E The section of *Kaddish D'Rabanan* that makes it unique (relative to other versions of the *Kaddish*) asks for blessing on the teachers, their students, and the students of their students. Why do you think the Rabbis created this blessing? What have you learned from another that you will pass on to someone else, creating a שַׁלְשֶׁלֶת הַקַּבָּלָה *shalshelet hakabbalah* — chain of tradition from teacher to student?

פְּסוּקֵי דְזִמְרָה

P'SUKEI D'ZIMRAH— VERSES OF PRAISE

בָּרוּךְ שֶׁאָמַר וְהָיָה הָעוֹלָם,
בָּרוּךְ הוּא.
בָּרוּךְ עוֹשֶׂה בְרֵאשִׁית,
בָּרוּךְ אוֹמֵר וְעוֹשֶׂה,
בָּרוּךְ גּוֹזֵר וּמְקַיֵּם,
בָּרוּךְ מְרַחֵם עַל הָאָרֶץ,
בָּרוּךְ מְרַחֵם עַל הַבְּרִיּוֹת,
בָּרוּךְ מְשַׁלֵּם שָׂכָר טוֹב לִירֵאָיו.
בָּרוּךְ חַי לָעַד וְקַיָּם לָנֶצַח.
בָּרוּךְ פּוֹדֶה וּמַצִּיל, בָּרוּךְ שְׁמוֹ.
בִּשְׁבָחוֹת וּבִזְמִרוֹת נְגַדֶּלְךָ
וּנְשַׁבֵּחֲךָ וּנְפָאֶרְךָ וְנַזְכִּיר
שִׁמְךָ וְנַמְלִיכְךָ, מַלְכֵּנוּ אֱלֹהֵינוּ.
יָחִיד, חֵי הָעוֹלָמִים, מֶלֶךְ מְשֻׁבָּח
וּמְפֹאָר, עֲדֵי עַד שְׁמוֹ הַגָּדוֹל.
בָּרוּךְ אַתָּה, יְיָ,
מֶלֶךְ מְהֻלָּל בַּתִּשְׁבָּחוֹת.

BLESSED is the One who spoke and the world came to be. Blessed is the One! Blessed is the One who continually authors creation. Blessed is the One whose word is deed; blessed is the One who decrees and fulfills. Blessed is the One who is compassionate towards the world; blessed is the One who is compassionate towards all creatures. Blessed is the One who rewards the reverent; blessed is the One who exists for all time, ever-enduring. Blessed is the One who redeems and saves; blessed is God's Name! With songs of praise, we extol You and proclaim Your Sovereignty, for You are the Source of life in the universe. One God, Life of the Universe, praised and glorious Ruler, Your Name is Eternal.

Blessed are You, Adonai, Sovereign who is glorified through praise.

בָּרוּךְ אַתָּה, יְיָ, מֶלֶךְ מְהֻלָּל בַּתִּשְׁבָּחוֹת.

פְּסוּקֵי דְזִמְרָה *P'sukei D'zimrah . . . Verses of Praise,* might be viewed as "prayer before prayer." It functions as the warm-up for the morning service, a recognition that prayerfulness cannot be summoned on demand. *Lawrence Hoffman*

PSALM 100:1-5 — A PSALM OF GRATITUDE

Omitted on Chol HaMo-eid Pesach

מִזְמוֹר לְתוֹדָה.
הָרִיעוּ לַייָ כָּל־הָאָרֶץ.
עִבְדוּ אֶת־יְיָ בְּשִׂמְחָה,
בֹּאוּ לְפָנָיו בִּרְנָנָה.
דְּעוּ כִּי יְיָ הוּא אֱלֹהִים,
הוּא עָשָׂנוּ, וְלוֹ אֲנַחְנוּ,
עַמּוֹ וְצֹאן מַרְעִיתוֹ.
בֹּאוּ שְׁעָרָיו בְּתוֹדָה,
חֲצֵרֹתָיו בִּתְהִלָּה.
הוֹדוּ לוֹ, בָּרְכוּ שְׁמוֹ.
כִּי־טוֹב יְיָ, לְעוֹלָם חַסְדּוֹ,
וְעַד־דֹּר וָדֹר אֱמוּנָתוֹ.

A PSALM OF PRAISE.
Raise a shout for Adonai, all the earth;
worship Adonai in gladness;
come into God's presence with shouts of joy.
Acknowledge that Adonai is God;
God made us and we are God's,
God's people, the flock God tends.
Enter God's gates with praise,
God's courts with acclamation.
Praise God!
Bless God's name!
For Adonai is good;
God's steadfast love is eternal;
God's faithfulness is for all generations.

בָּרוּךְ שֶׁאָמַר וְהָיָה הָעוֹלָם *Baruch she-amar v'hayah haolam . . . Blessed is the One who spoke and the world came to be . . .* The first mystery is simply that there is a mystery, one that can never be explained or understood. The Hebrew word for *universe,* עוֹלָם *olam,* comes from the word for *hidden.* Something of the Holy One is hidden within. *Lawrence Kushner*

בָּרוּךְ שֶׁאָמַר

מִזְמוֹר ק'

אַשְׁרֵי

מִזְמוֹר קמ"ה

מִזְמוֹר ק"נ

יִשְׁתַּבַּח

חֲצִי קַדִּישׁ

אַשְׁרֵי יוֹשְׁבֵי בֵיתֶךָ,
עוֹד יְהַלְלוּךָ סֶּלָה.
אַשְׁרֵי הָעָם שֶׁכָּכָה לּוֹ,
אַשְׁרֵי הָעָם שֶׁיְיָ אֱלֹהָיו.

HAPPY are those who dwell in Your house; they forever praise You!
Happy the people who have it so; happy the people whose God is Adonai.

DAVID'S PSALM ⸺ PSALM 145:1-21

תְּהִלָּה לְדָוִד.
אֲרוֹמִמְךָ אֱלוֹהַי הַמֶּלֶךְ,
וַאֲבָרְכָה שִׁמְךָ לְעוֹלָם וָעֶד.

בְּכָל־יוֹם אֲבָרְכֶךָּ,
וַאֲהַלְלָה שִׁמְךָ לְעוֹלָם וָעֶד.

גָּדוֹל יְיָ וּמְהֻלָּל מְאֹד,
וְלִגְדֻלָּתוֹ אֵין חֵקֶר.

דּוֹר לְדוֹר יְשַׁבַּח מַעֲשֶׂיךָ,
וּגְבוּרֹתֶיךָ יַגִּידוּ.

הֲדַר כְּבוֹד הוֹדֶךָ,
וְדִבְרֵי נִפְלְאֹתֶיךָ אָשִׂיחָה.

וֶעֱזוּז נוֹרְאֹתֶיךָ יֹאמֵרוּ,
וּגְדֻלָּתְךָ אֲסַפְּרֶנָּה.

זֵכֶר רַב־טוּבְךָ יַבִּיעוּ,
וְצִדְקָתְךָ יְרַנֵּנוּ.

חַנּוּן וְרַחוּם יְיָ,
אֶרֶךְ אַפַּיִם וּגְדָל־חָסֶד.

טוֹב יְיָ לַכֹּל,
וְרַחֲמָיו עַל־כָּל־מַעֲשָׂיו.

יוֹדוּךָ יְיָ כָּל־מַעֲשֶׂיךָ,
וַחֲסִידֶיךָ יְבָרְכוּכָה.

אַשְׁרֵי יוֹשְׁבֵי *Ashrei yoshvei . . . Happy are those who dwell . . .* Psalm 84:5
אַשְׁרֵי הָעָם *Ashrei haam . . . Happy the people . . .* Psalm 144:15

כְּבוֹד מַלְכוּתְךָ יֹאמֵרוּ,
וּגְבוּרָתְךָ יְדַבֵּרוּ.

לְהוֹדִיעַ לִבְנֵי הָאָדָם גְּבוּרֹתָיו,
וּכְבוֹד הֲדַר מַלְכוּתוֹ.

מַלְכוּתְךָ מַלְכוּת כָּל־עֹלָמִים,
וּמֶמְשַׁלְתְּךָ בְּכָל־דֹּר וָדֹר.

סוֹמֵךְ יְיָ לְכָל־הַנֹּפְלִים,
וְזוֹקֵף לְכָל־הַכְּפוּפִים.

עֵינֵי כֹל אֵלֶיךָ יְשַׂבֵּרוּ,
וְאַתָּה נוֹתֵן־לָהֶם אֶת־אָכְלָם בְּעִתּוֹ.

פּוֹתֵחַ אֶת־יָדֶךָ,
וּמַשְׂבִּיעַ לְכָל־חַי רָצוֹן.

צַדִּיק יְיָ בְּכָל־דְּרָכָיו,
וְחָסִיד בְּכָל־מַעֲשָׂיו.

קָרוֹב יְיָ לְכָל־קֹרְאָיו,
לְכֹל אֲשֶׁר יִקְרָאֻהוּ בֶאֱמֶת.

רְצוֹן־יְרֵאָיו יַעֲשֶׂה,
וְאֶת־שַׁוְעָתָם יִשְׁמַע וְיוֹשִׁיעֵם.

שׁוֹמֵר יְיָ אֶת־כָּל־אֹהֲבָיו,
וְאֵת כָּל־הָרְשָׁעִים יַשְׁמִיד.

תְּהִלַּת יְיָ יְדַבֶּר־פִּי,
וִיבָרֵךְ כָּל־בָּשָׂר שֵׁם קָדְשׁוֹ
לְעוֹלָם וָעֶד.

וַאֲנַחְנוּ נְבָרֵךְ יָהּ
מֵעַתָּה וְעַד עוֹלָם, הַלְלוּיָהּ.

וַאֲנַחְנוּ נְבָרֵךְ יָהּ *Vaanachnu n'vareich Yah . . . and all creatures shall bless . . .* Psalm 115:18

בָּרוּךְ שֶׁאָמַר

מִזְמוֹר ק'

אַשְׁרֵי

מִזְמוֹר קמ"ה

מִזְמוֹר ק"נ

יִשְׁתַּבַּח

חֲצִי קַדִּישׁ

Baruch She-amar
Psalm 100
Ashrei
Psalm 145
Psalm 150
Yishtabach
Chatzi Kaddish

DAVID'S SONG OF PRAISE

I will extol You, my God and sovereign,
and bless Your name forever and ever.

Every day will I bless You and praise Your name forever and ever.
Great is Adonai and much acclaimed;
God's greatness cannot be fathomed.

One generation shall laud Your works to another
and declare Your mighty acts.
The glorious majesty of Your splendor
and Your wondrous acts will I recite.

They shall talk of the might of Your awesome deeds,
and I will recount Your greatness.
They shall celebrate Your abundant goodness,
and sing joyously of Your beneficence.

Adonai is gracious and compassionate,
slow to anger and abounding in kindness.
Adonai is good to all, and God's mercy is upon all God's works.

All Your works shall praise You, Adonai,
and Your faithful ones shall bless You.
They shall talk of the majesty of Your sovereignty,
and speak of Your might,

to make God's mighty acts known among mortals
and the majestic glory of Your sovereignty.
Your sovereignty is eternal; Your dominion is for all generations.

Adonai supports all who stumble,
and makes all who are bent stand straight.
The eyes of all look to You expectantly,
and You give them their food when it is due.

You give it openhandedly,
feeding every creature to its heart's content.
Adonai is beneficent in all ways and faithful in all works.

Adonai is near to all who call
to all who call upon God with sincerity.
Adonai fulfills the wishes of those who fear God;
Adonai hears their cry and delivers them.

Adonai watches over all who love God,
but all the wicked God will destroy.
My mouth shall utter the praise of Adonai,
and all creatures shall bless God's holy name forever and ever.

We will bless You God now and always.
Hallelujah!

בָּרוּךְ שֶׁאָמַר

מִזְמוֹר ק'

אַשְׁרֵי

מִזְמוֹר קמ"ה

מִזְמוֹר ק"ג

יִשְׁתַּבַּח

חֲצִי קַדִּישׁ

PSALM 150:1–6

הַלְלוּ יָהּ!
הַלְלוּ־אֵל בְּקָדְשׁוֹ,
הַלְלוּהוּ בִּרְקִיעַ עֻזּוֹ.
הַלְלוּהוּ בִּגְבוּרֹתָיו,
הַלְלוּהוּ כְּרֹב גֻּדְלוֹ.
הַלְלוּהוּ בְּתֵקַע שׁוֹפָר,
הַלְלוּהוּ בְּנֵבֶל וְכִנּוֹר.
הַלְלוּהוּ בְּתֹף וּמָחוֹל,
הַלְלוּהוּ בְּמִנִּים וְעוּגָב.
הַלְלוּהוּ בְּצִלְצְלֵי־שָׁמַע,
הַלְלוּהוּ בְּצִלְצְלֵי־תְרוּעָה.
כֹּל הַנְּשָׁמָה תְּהַלֵּל יָהּ,
הַלְלוּ־יָהּ!

HALLELUJAH!
Praise God in God's sanctuary;
praise God in the sky, God's stronghold.
Praise God for mighty acts;
praise God for God's exceeding greatness.
Praise God with blasts of the horn;
praise God with harp and lyre.
Praise God with timbrel and dance;
praise God with lute and pipe.
Praise God with resounding cymbals;
praise God with loud-clashing cymbals.
Let all that breathes praise Adonai.
Hallelujah!

Every praise could be music, the voice singing in harmony with the universe and its Creator!

יִשְׁתַּבַּח שִׁמְךָ לָעַד מַלְכֵּנוּ,
הָאֵל הַמֶּלֶךְ הַגָּדוֹל וְהַקָּדוֹשׁ
בַּשָּׁמַיִם וּבָאָרֶץ.
כִּי לְךָ נָאֶה, יְיָ אֱלֹהֵינוּ
וֵאלֹהֵי אֲבוֹתֵינוּ וְאִמּוֹתֵינוּ,
שִׁיר וּשְׁבָחָה, הַלֵּל וְזִמְרָה,
עֹז וּמֶמְשָׁלָה, נֶצַח
גְּדֻלָּה וּגְבוּרָה,
תְּהִלָּה וְתִפְאֶרֶת, קְדֻשָּׁה וּמַלְכוּת,
בְּרָכוֹת וְהוֹדָאוֹת מֵעַתָּה וְעַד עוֹלָם.

YOU SHALL ALWAYS BE PRAISED,
great and holy God, our Sovereign in heaven and on earth.
Songs of praise and psalms of adoration become You,
acknowledging Your might and Your dominion.
Yours are strength and sovereignty, sanctity, grandeur and glory always.
We offer our devotion, open our hearts in acclamation.

בָּרוּךְ אַתָּה, יְיָ,
אֵל מֶלֶךְ גָּדוֹל בַּתִּשְׁבָּחוֹת,
אֵל הַהוֹדָאוֹת, אֲדוֹן הַנִּפְלָאוֹת,
הַבּוֹחֵר בְּשִׁירֵי זִמְרָה,
מֶלֶךְ אֵל חֵי הָעוֹלָמִים.

Praised are You, Sovereign of wonders,
crowned in adoration,
delighting in song,
Eternal Majesty.

The help we really need is spiritual. What we really need is courage, patience, strength, wisdom, and serenity. These can only come to us from the realm of the spirit, from that source of the spirit that we call God . . . but strength and courage, fortitude and patience and hope are not tangible and have no sound. They have no sound; they have no touch. Yet when you need them they are more real and more true and more important than anything tangible could ever be. *Harold Hahn*

יִתְגַּדַּל וְיִתְקַדַּשׁ שְׁמֵהּ רַבָּא
בְּעָלְמָא דִּי בְרָא כִרְעוּתֵהּ,
וְיַמְלִיךְ מַלְכוּתֵהּ
בְּחַיֵּיכוֹן וּבְיוֹמֵיכוֹן
וּבְחַיֵּי דְכָל בֵּית יִשְׂרָאֵל,
בַּעֲגָלָא וּבִזְמַן קָרִיב,
וְאִמְרוּ: אָמֵן.

יְהֵא שְׁמֵהּ רַבָּא מְבָרַךְ
לְעָלַם וּלְעָלְמֵי עָלְמַיָּא.

יִתְבָּרַךְ וְיִשְׁתַּבַּח וְיִתְפָּאַר
וְיִתְרוֹמַם וְיִתְנַשֵּׂא,
וְיִתְהַדַּר וְיִתְעַלֶּה וְיִתְהַלַּל
שְׁמֵהּ דְּקֻדְשָׁא בְּרִיךְ הוּא,
לְעֵלָּא מִן כָּל בִּרְכָתָא וְשִׁירָתָא,
תֻּשְׁבְּחָתָא וְנֶחֱמָתָא,
דַּאֲמִירָן בְּעָלְמָא, וְאִמְרוּ: אָמֵן.

EXALTED and hallowed be God's great name,
in the world which God created, according to plan.
May God's majesty be revealed in the days of our lifetime
and the life of all Israel — speedily, imminently.
To which we say: Amen.

Blessed be God's great name to all eternity.

Blessed, praised, honored, exalted,
extolled, glorified, adored, and lauded
be the name of the Holy Blessed One,
beyond all earthly words and songs of blessing, praise, and comfort.
To which we say: Amen.

Better a few prayers spoken with intention than many words prayed without intention.
Shulchan Aruch Orach Chayim 1:4

As we begin the two main sections of our prayer service, we consider: Rabbi Shimon says, Take care when reciting the שְׁמַע *Sh'ma* and the עֲמִידָה *Amidah*. And when you pray, do not make your prayer rote. Rather, it should be filled with mercy and confession before the Almighty, as it is said, "... For God is gracious and compassionate slow to anger, abounding in kindness.... (Joel 2:13)"
Pirkei Avot 2:13

שְׁמַע וּבִרְכוֹתֶיהָ

SH'MA UVIRCHOTEHA — SH'MA AND ITS BLESSINGS

בָּרְכוּ אֶת יְיָ הַמְבֹרָךְ!
בָּרוּךְ יְיָ הַמְבֹרָךְ
לְעוֹלָם וָעֶד!

PRAISE ADONAI to whom praise is due forever!
Praised be Adonai to whom praise is due,
now and forever!

All holy acts require summoning.

For those who choose: The prayer leader at the word בָּרְכוּ *Bar'chu* (the call to worship) bends the knees and bows from the waist, and at יְיָ *Adonai* stands straight. בָּרוּךְ יְיָ *Baruch Adonai* is the communal response, whereupon the community repeats the choreography of the first line.

Bar'chu
Yotzeir
Ahavah Rabbah
Sh'ma
V'ahavta
L'maan Tizk'ru
Vayomer Adonai
Emet v'Yatziv
Mi Chamochah

בָּרְכוּ
יוֹצֵר
אַהֲבָה רַבָּה
שְׁמַע
וְאָהַבְתָּ
לְמַעַן תִּזְכְּרוּ
וַיֹּאמֶר יְיָ
אֱמֶת וְיַצִּיב
מִי־כָמֹכָה

בָּרוּךְ אַתָּה, יְיָ
אֱלֹהֵינוּ, מֶלֶךְ הָעוֹלָם,
יוֹצֵר אוֹר וּבוֹרֵא חֹשֶׁךְ,
עֹשֶׂה שָׁלוֹם וּבוֹרֵא אֶת־הַכֹּל.
הַמֵּאִיר לָאָרֶץ
וְלַדָּרִים עָלֶיהָ בְּרַחֲמִים,
וּבְטוּבוֹ מְחַדֵּשׁ בְּכָל יוֹם תָּמִיד
מַעֲשֵׂה בְרֵאשִׁית.
מָה רַבּוּ מַעֲשֶׂיךָ, יְיָ,
כֻּלָּם בְּחָכְמָה עָשִׂיתָ,
מָלְאָה הָאָרֶץ קִנְיָנֶךָ.
תִּתְבָּרַךְ, יְיָ אֱלֹהֵינוּ,
עַל שֶׁבַח מַעֲשֵׂה יָדֶיךָ
וְעַל מְאוֹרֵי אוֹר שֶׁעָשִׂיתָ,
יְפָאֲרוּךָ סֶּלָה.
אוֹר חָדָשׁ עַל צִיּוֹן תָּאִיר,
וְנִזְכֶּה כֻלָּנוּ מְהֵרָה לְאוֹרוֹ.
בָּרוּךְ אַתָּה, יְיָ, יוֹצֵר הַמְּאוֹרוֹת.

PRAISED ARE YOU, Adonai our God, Sovereign of the universe,
Creator of light and darkness, who makes peace and fashions all things.
In mercy, You illumine the world and those who live upon it.
In Your goodness You daily renew creation.
How numerous are Your works, Adonai!
In wisdom, You formed them all, filling the earth with Your creatures.
Be praised, Adonai our God, for the excellent work of Your hands,
and for the lights You created; may they glorify You.
Shine a new light upon Zion, that we all may swiftly merit its radiance.
Praised are You, Adonai, Creator of all heavenly lights.

בָּרוּךְ אַתָּה, יְיָ, יוֹצֵר הַמְּאוֹרוֹת.

אוֹר חָדָשׁ עַל צִיּוֹן תָּאִיר *Or chadash al Tzion ta-ir . . . Shine a new light upon Zion . . .* Classical Reform prayerbook authors in the Diaspora consistently omitted this line with its mention of Zion from the liturgy because of their opposition to Jewish nationalism. With the restoration of this passage to *Mishkan T'filah*, our movement consciously affirms its devotion to the modern State of Israel and signals its recognition of the religious significance of the reborn Jewish commonwealth. *David Ellenson*

בָּרוּךְ אַתָּה, יְיָ . . . יוֹצֵר אוֹר וּבוֹרֵא חֹשֶׁךְ *Baruch atah, Adonai . . . yotzeir or uvorei choshech . . . Praised are You, Adonai . . . Creator of light and darkness . . .* based on Isaiah 45:7

מָה רַבּוּ *Mah rabu . . . How numerous . . .* Psalm 104:24

I With your eyes closed, turn toward the light. Feel the various sources of light — natural and artificial. Does one give more strength than the other? Does the sunrise, after a long night, recharge and reinvigorate? How does the light affect you?

בָּרְכוּ
יוֹצֵר
אַהֲבָה רַבָּה
שְׁמַע
וְאָהַבְתָּ
לְמַעַן תִּזְכְּרוּ
וַיֹּאמֶר יְיָ
אֱמֶת וְיַצִּיב
מִי־כָמְכָה

אַהֲבָה רַבָּה אֲהַבְתָּנוּ, יְיָ אֱלֹהֵינוּ,
חֶמְלָה גְדוֹלָה וִיתֵרָה חָמַלְתָּ עָלֵינוּ.
בַּעֲבוּר אֲבוֹתֵינוּ וְאִמּוֹתֵינוּ שֶׁבָּטְחוּ בְךָ
וַתְּלַמְּדֵם חֻקֵּי חַיִּים, כֵּן תְּחָנֵּנוּ
וּתְלַמְּדֵנוּ. הַמְרַחֵם, רַחֵם עָלֵינוּ,
וְתֵן בְּלִבֵּנוּ לְהָבִין וּלְהַשְׂכִּיל, לִשְׁמֹעַ,
לִלְמֹד וּלְלַמֵּד, לִשְׁמֹר וְלַעֲשׂוֹת וּלְקַיֵּם
אֶת־כָּל־דִּבְרֵי תַלְמוּד תּוֹרָתֶךָ בְּאַהֲבָה.

HOW DEEPLY You have loved us Adonai, our God, gracing us with surpassing compassion! On account of our forebears whose trust led You to teach them the laws of life, be gracious to us, teaching us as well. O Merciful One, have mercy on us by making us able to understand and discern, to heed, learn, and teach, and, lovingly, to observe, perform, and fulfill all that is in Your Torah.

וְהָאֵר עֵינֵינוּ בְּתוֹרָתֶךָ,
וְדַבֵּק לִבֵּנוּ בְּמִצְוֹתֶיךָ,
וְיַחֵד לְבָבֵנוּ לְאַהֲבָה
וּלְיִרְאָה אֶת־שְׁמֶךָ,
וְלֹא נֵבוֹשׁ וְלֹא נִכָּלֵם,
וְלֹא נִכָּשֵׁל לְעוֹלָם וָעֶד.
כִּי בְשֵׁם קָדְשְׁךָ הַגָּדוֹל וְהַנּוֹרָא
בָּטָחְנוּ, נָגִילָה וְנִשְׂמְחָה בִּישׁוּעָתֶךָ.
וַהֲבִיאֵנוּ לְשָׁלוֹם מֵאַרְבַּע כַּנְפוֹת
הָאָרֶץ, וְתוֹלִיכֵנוּ קוֹמְמִיּוּת לְאַרְצֵנוּ.
כִּי אֵל פּוֹעֵל יְשׁוּעוֹת אָתָּה, וּבָנוּ בָחַרְתָּ
וְקֵרַבְתָּנוּ לְשִׁמְךָ הַגָּדוֹל סֶלָה בֶּאֱמֶת,
לְהוֹדוֹת לְךָ וּלְיַחֶדְךָ בְּאַהֲבָה.
בָּרוּךְ אַתָּה, יְיָ,
הַבּוֹחֵר בְּעַמּוֹ יִשְׂרָאֵל בְּאַהֲבָה.

Enlighten our eyes with Your Torah, focus our minds on Your mitzvot, unite our hearts in love and reverence for Your Name. Then we will never feel shame, never deserve rebuke, and never stumble. Having trusted in Your great and awesome holiness, we shall celebrate Your salvation with joy.

Gather us in peace from the four corners of the earth and lead us upright to our land.
For You, O God, work wonders. You chose us. Truly, you drew us near to Your
Great Name, that we might acknowledge You, declaring You One in love.

Praised be You, Adonai, who chooses Your people Israel in love.

בָּרוּךְ אַתָּה, יְיָ, הַבּוֹחֵר בְּעַמּוֹ יִשְׂרָאֵל בְּאַהֲבָה.

E Each of us longs for unconditional love — for the comfort and security of knowing that no matter what we do, we are cherished. However, we often feel an obligation to give back to those who give to us. What is our responsibility toward those who give us love? What if that love is coming from God?

For those who choose: At the words וַהֲבִיאֵנוּ לְשָׁלוֹם *Vahavi-einu l'shalom, Gather us in peace,* one gathers the four fringes of the tallit in the left hand and holds them throughout the שְׁמַע *Sh'ma* to symbolize the ingathering of our people.

שְׁמַע יִשְׂרָאֵל יְהוָה אֱלֹהֵינוּ יְהוָה אֶחָד!

Hear, O Israel, Adonai is our God, Adonai is One!

שְׁמַע יִשְׂרָאֵל *Sh'ma Yisrael . . . Hear, O Israel . . .* Deuteronomy 6:4

בָּרוּךְ שֵׁם כְּבוֹד מַלְכוּתוֹ לְעוֹלָם וָעֶד.

Blessed is God's glorious majesty forever and ever.

The enlarged ע *ayin* at the end of שמע *Sh'ma* (Hear) and the enlarged ד *dalet* at the end of אחד *echad* (one) combine to spell עד *eid* (witness). We recite the *Sh'ma* to bear witness to the Oneness of God.

בָּרוּךְ שֵׁם כְּבוֹד *Baruch shem k'vod . . . Blessed is God's glorious . . . M. Yoma 3:8, inspired by* Nehemiah 9:5

בָּרְכוּ

יוֹצֵר

אַהֲבָה רַבָּה

שְׁמַע

וְאָהַבְתָּ

לְמַעַן תִּזְכְּרוּ

וַיֹּאמֶר יְיָ

אֱמֶת וְיַצִּיב

מִי־כָמֹכָה

וְאָהַבְתָּ אֵת יְיָ אֱלֹהֶיךָ
בְּכָל־לְבָבְךָ וּבְכָל־נַפְשְׁךָ וּבְכָל־
מְאֹדֶךָ: וְהָיוּ הַדְּבָרִים הָאֵלֶּה
אֲשֶׁר אָנֹכִי מְצַוְּךָ הַיּוֹם עַל־
לְבָבֶךָ: וְשִׁנַּנְתָּם לְבָנֶיךָ וְדִבַּרְתָּ
בָּם בְּשִׁבְתְּךָ בְּבֵיתֶךָ וּבְלֶכְתְּךָ
בַדֶּרֶךְ וּבְשָׁכְבְּךָ וּבְקוּמֶךָ:
וּקְשַׁרְתָּם לְאוֹת עַל־יָדֶךָ וְהָיוּ
לְטֹטָפֹת בֵּין עֵינֶיךָ: וּכְתַבְתָּם
עַל־מְזֻזוֹת בֵּיתֶךָ וּבִשְׁעָרֶיךָ:

YOU SHALL LOVE Adonai your God with all your heart,
with all your soul, and with all your might.
Take to heart these instructions with which I charge you this day.
Impress them upon your children.
Recite them when you stay at home and when you are away,
when you lie down and when you get up.
Bind them as a sign on your hand and let them serve as a symbol on your forehead;
inscribe them on the doorposts of your house and on your gates.

Continue or turn to pages 68–69.

לְמַעַן תִּזְכְּרוּ וַעֲשִׂיתֶם אֶת־
כָּל־מִצְוֹתָי וִהְיִיתֶם קְדֹשִׁים
לֵאלֹהֵיכֶם: אֲנִי יְיָ אֱלֹהֵיכֶם אֲשֶׁר
הוֹצֵאתִי אֶתְכֶם מֵאֶרֶץ
מִצְרַיִם לִהְיוֹת לָכֶם לֵאלֹהִים
אֲנִי יְיָ אֱלֹהֵיכֶם:

Thus you shall remember to observe all My commandments
and to be holy to your God.
I am Adonai, your God, who brought you out of the land of Egypt to be your God:
I am Adonai your God.

יְיָ אֱלֹהֵיכֶם אֱמֶת.

Turn to pages 70–71.

For those who choose: At the end of the שְׁמַע *Sh'ma*, after the words יְיָ אֱלֹהֵיכֶם *Adonai Eloheichem,* the word אֱמֶת *emet* ("true") is added as an immediate affirmation of its truth.

וְאָהַבְתָּ *V'ahavta . . . You shall love . . .* Deuteronomy 6:5–9

לְמַעַן תִּזְכְּרוּ *L'maan tizk'ru . . . Thus you shall remember . . .* Numbers 15:40–41

בָּרְכוּ
יוֹצֵר
אַהֲבָה רַבָּה
שְׁמַע
וְאָהַבְתָּ
לְמַעַן תִּזְכְּרוּ
וַיֹּאמֶר יְיָ
אֱמֶת וְיַצִּיב
מִי־כָמֹכָה

וַיֹּאמֶר יְיָ אֶל־מֹשֶׁה לֵּאמֹר׃
דַּבֵּר אֶל־בְּנֵי יִשְׂרָאֵל
וְאָמַרְתָּ אֲלֵהֶם וְעָשׂוּ לָהֶם צִיצִת
עַל־כַּנְפֵי בִגְדֵיהֶם לְדֹרֹתָם
וְנָתְנוּ עַל־צִיצִת הַכָּנָף פְּתִיל תְּכֵלֶת׃
וְהָיָה לָכֶם לְצִיצִת וּרְאִיתֶם אֹתוֹ
וּזְכַרְתֶּם אֶת־כָּל־מִצְוֹת יְיָ
וַעֲשִׂיתֶם אֹתָם
וְלֹא תָתוּרוּ אַחֲרֵי לְבַבְכֶם
וְאַחֲרֵי עֵינֵיכֶם
אֲשֶׁר־אַתֶּם זֹנִים אַחֲרֵיהֶם׃

ADONAI SAID TO MOSES as follows:
Speak to the Israelite people and instruct them to make for themselves
fringes on the corners of their garments throughout the ages;
let them attach a cord of blue to the fringe at each corner.
That shall be your fringe; look at it and recall all the commandments of Adonai
and observe them, so that you do not follow your heart and eyes in your lustful urge.

לְמַעַן תִּזְכְּרוּ וַעֲשִׂיתֶם אֶת־
כָּל־מִצְוֹתָי וִהְיִיתֶם קְדֹשִׁים
לֵאלֹהֵיכֶם׃ אֲנִי יְיָ אֱלֹהֵיכֶם אֲשֶׁר
הוֹצֵאתִי אֶתְכֶם מֵאֶרֶץ
מִצְרַיִם לִהְיוֹת לָכֶם לֵאלֹהִים
אֲנִי יְיָ אֱלֹהֵיכֶם׃

Thus you shall remember to observe all My commandments
and to be holy to your God.
I am Adonai, your God, who brought you out of the land of Egypt to be your God:
I am Adonai your God.

יְיָ אֱלֹהֵיכֶם אֱמֶת.

For those who choose: At the word צִיצִת *tzitzit* ("fringes") and at the final word אֱמֶת *emet* ("truth"), one brings the *tzitzit* to one's lips.

וַיֹּאמֶר יְיָ אֶל־מֹשֶׁה *Vayomer Adonai el Moshe . . . Adonai said to Moses . . .* Numbers 15:37–39

לְמַעַן תִּזְכְּרוּ *L'maan tizk'ru . . . Thus you shall remember . . .* Numbers 15:40–41

E Wearing a tallit is meant to be a reminder of קַבָּלַת עוֹל הַמִּצְווֹת *kabbalat ol hamitzvot* — accepting the responsibility of the mitzvot. Many people wear one for the first time when they become bar or bat mitzvah. Have you ever worn a tallit? How do you interpret the symbolism of wearing the garment for the first time on becoming bar or bat mitzvah?

אֱמֶת וְיַצִּיב וְאָהוּב
וְחָבִיב וְנוֹרָא וְאַדִּיר וְטוֹב וְיָפֶה
הַדָּבָר הַזֶּה עָלֵינוּ לְעוֹלָם וָעֶד.
אֱמֶת, אֱלֹהֵי עוֹלָם מַלְכֵּנוּ,
צוּר יַעֲקֹב, מָגֵן יִשְׁעֵנוּ.
לְדֹר וָדֹר הוּא קַיָּם
וּשְׁמוֹ קַיָּם וְכִסְאוֹ נָכוֹן
וּמַלְכוּתוֹ וֶאֱמוּנָתוֹ לָעַד קַיָּמֶת.
וּדְבָרָיו חָיִים וְקַיָּמִים,
נֶאֱמָנִים וְנֶחֱמָדִים
לָעַד וּלְעוֹלְמֵי עוֹלָמִים.
מִמִּצְרַיִם גְּאַלְתָּנוּ, יְיָ אֱלֹהֵינוּ,
וּמִבֵּית עֲבָדִים פְּדִיתָנוּ. עַל זֹאת
שִׁבְּחוּ אֲהוּבִים וְרוֹמְמוּ אֵל,
וְנָתְנוּ יְדִידִים זְמִירוֹת שִׁירוֹת
וְתִשְׁבָּחוֹת, בְּרָכוֹת וְהוֹדָאוֹת
לְמֶלֶךְ אֵל חַי וְקַיָּם. רָם
וְנִשָּׂא, גָּדוֹל וְנוֹרָא, מַשְׁפִּיל
גֵּאִים וּמַגְבִּיהַּ שְׁפָלִים, מוֹצִיא
אֲסִירִים וּפוֹדֶה עֲנָוִים וְעוֹזֵר
דַּלִּים וְעוֹנֶה לְעַמּוֹ בְּעֵת שַׁוְּעָם.
תְּהִלּוֹת לְאֵל עֶלְיוֹן, בָּרוּךְ
הוּא וּמְבֹרָךְ. מֹשֶׁה וּמִרְיָם
וּבְנֵי יִשְׂרָאֵל לְךָ עָנוּ שִׁירָה
בְּשִׂמְחָה רַבָּה וְאָמְרוּ כֻלָּם:

FOR US, this eternal teaching is true and enduring, beloved and precious, awesome, good and beautiful. The God of the universe is truly our Sovereign, the Rock of Jacob, our Protecting Shield. God endures through all generations; God's name persists; God's throne is firm; God's sovereignty and faithfulness last forever. God's words live and endure, faithful and precious for eternity.

From Egypt You redeemed us, freeing us from bondage. For that, Your beloved sang praise, exalting You. Your dear ones offered hymns, songs, praise, blessing, and thanksgiving to You as Sovereign, the living and enduring God. High and exalted, great and awesome, God ever humbles the proud, raises the lowly, frees the imprisoned, redeems the afflicted, helps the oppressed, answering our people when we cry out. Praise to God Most High; blessed is God and deserving of blessing! In great joy, Moses, Miriam and Israel responded with song to You, all of them proclaiming:

E What teaching have you received that is "true and enduring, beloved and precious, awesome, good, and beautiful"? How or from whom did you learn it?

בָּרְכוּ
יוֹצֵר
אַהֲבָה רַבָּה
שְׁמַע
וְאָהַבְתָּ
לְמַעַן תִּזְכְּרוּ
וַיֹּאמֶר יְיָ
אֱמֶת וְיַצִּיב
מִי־כָמֹכָה

מִי־כָמֹכָה בָּאֵלִם, יְיָ!
מִי כָּמֹכָה נֶאְדָּר בַּקֹּדֶשׁ,
נוֹרָא תְהִלֹּת, עֹשֵׂה פֶלֶא!

שִׁירָה חֲדָשָׁה שִׁבְּחוּ גְאוּלִים
לְשִׁמְךָ עַל שְׂפַת הַיָּם.
יַחַד כֻּלָּם הוֹדוּ וְהִמְלִיכוּ וְאָמְרוּ:
יְיָ יִמְלֹךְ לְעוֹלָם וָעֶד.

צוּר יִשְׂרָאֵל, קוּמָה בְּעֶזְרַת יִשְׂרָאֵל
וּפְדֵה כִנְאֻמֶךָ יְהוּדָה וְיִשְׂרָאֵל.
גֹּאֲלֵנוּ יְיָ צְבָאוֹת שְׁמוֹ,
קְדוֹשׁ יִשְׂרָאֵל.
בָּרוּךְ אַתָּה, יְיָ, גָּאַל יִשְׂרָאֵל.

WHO IS LIKE YOU, O God,
among the gods that are worshipped?
Who is like You, majestic in holiness,
awesome in splendor, working wonders?

With new song, inspired,
at the shore of the Sea, the redeemed sang Your praise.
In unison they all offered thanks.
Acknowledging Your Sovereignty, they said:
"Adonai will reign forever!"

Rock of Israel, rise in support of Israel
and redeem Judah and Israel as You promised.
Our redeemer, *Adonai Tz'vaot* is Your Name.
Blessed are You, Adonai, who redeems Israel.

בָּרוּךְ אַתָּה, יְיָ, גָּאַל יִשְׂרָאֵל.

For those who choose: When the prayer leader recites the word קוּמָה *kumah* ("rise"), the congregation rises for the עֲמִידָה *Amidah, the Standing Prayer.*

מִי־כָמֹכָה *Mi chamochah . . . Who is like You? . . .* Exodus 15:11

יְיָ יִמְלֹךְ לְעֹלָם וָעֶד *Adonai yimloch l'olam va-ed . . . Adonai will reign forever . . .* Exodus 15:18

Bar'chu
Yotzeir
Ahavah Rabbah
Sh'ma
V'ahavta
L'maan Tizk'ru
Vayomer Adonai
Emet v'Yatziv
Mi Chamochah

P Freedom is sometimes more exciting in theory than in actuality. After the Song at the Sea was over, after the celebration, what did the Israelites do next? What are some of the challenges of freedom?

אָבוֹת וְאִמָּהוֹת
גְּבוּרוֹת
קְדֻשָּׁה

בִּינָה
תְּשׁוּבָה
סְלִיחָה
גְּאֻלָּה
רְפוּאָה
בִּרְכַּת הַשָּׁנִים
חֵרוּת
מִשְׁפָּט
עַל הָרִשְׁעָה
צַדִּיקִים
יְרוּשָׁלַיִם
יְשׁוּעָה
שׁוֹמֵעַ תְּפִלָּה

עֲבוֹדָה
הוֹדָאָה
שָׁלוֹם
תְּפִלַּת הַלֵּב

T'FILAH

אֲדֹנָי, שְׂפָתַי תִּפְתָּח,
וּפִי יַגִּיד תְּהִלָּתֶךָ.

ADONAI, open up my lips,
that my mouth may declare Your praise.

For those who choose: Before reciting the תְּפִלָּה *T'filah,* one takes three steps forward.

"Adonai, open my lips that my mouth may declare Your praise; for You have no delight in sacrifice. If I were to give a burnt offering, You would not be pleased." According to the Midrash, Israel said to God, "We are impoverished now that we cannot offer sacrifices." God answered, "I seek words from you now, as it is written, 'Take words with you when you return to your God' (Hosea 14:3)." *Midrash Sh'mot Rabbah 38:4*

אֲדֹנָי, שְׂפָתַי תִּפְתָּח *Adonai, s'fatai tiftach . . . Adonai, open up my lips . . .* Psalm 51:17

Weekday T'filah

Avot v'Imahot

G'vurot

K'dushah

Binah

T'shuvah

S'lichah

G'ulah

R'fuah

Birkat HaShanim

Cheirut

Mishpat

Al HaRishah

Tzadikim

Y'rushalayim

Y'shuah

Shomei-a T'filah

Avodah

Hodaah

Shalom

T'filat HaLev

בָּרוּךְ אַתָּה, יְיָ אֱלֹהֵינוּ
וֵאלֹהֵי אֲבוֹתֵינוּ וְאִמּוֹתֵינוּ, אֱלֹהֵי
אַבְרָהָם, אֱלֹהֵי יִצְחָק וֵאלֹהֵי יַעֲקֹב,
אֱלֹהֵי שָׂרָה, אֱלֹהֵי רִבְקָה, אֱלֹהֵי
רָחֵל וֵאלֹהֵי לֵאָה. הָאֵל הַגָּדוֹל
הַגִּבּוֹר וְהַנּוֹרָא, אֵל עֶלְיוֹן, גּוֹמֵל
חֲסָדִים טוֹבִים, וְקוֹנֵה הַכֹּל, וְזוֹכֵר
חַסְדֵי אָבוֹת וְאִמָּהוֹת, וּמֵבִיא גְאֻלָּה
לִבְנֵי בְנֵיהֶם לְמַעַן שְׁמוֹ בְּאַהֲבָה.

BETWEEN ROSH HASHANAH AND

זָכְרֵנוּ לְחַיִּים, — YOM KIPPUR
מֶלֶךְ חָפֵץ בַּחַיִּים,
וְכָתְבֵנוּ בְּסֵפֶר הַחַיִּים,
לְמַעַנְךָ אֱלֹהִים חַיִּים.

מֶלֶךְ עוֹזֵר וּמוֹשִׁיעַ וּמָגֵן.
בָּרוּךְ אַתָּה, יְיָ,
מָגֵן אַבְרָהָם וְעֶזְרַת שָׂרָה.

BLESSED ARE YOU, Adonai, our God,
God of our fathers and mothers,
God of Abraham, God of Isaac, and God of Jacob,
God of Sarah, God of Rebecca, God of Rachel, and God of Leah,
the great, mighty and awesome God, transcendent God
who bestows lovingkindness, creates everything out of love,
remembers the love of our fathers and mothers,
and brings redemption to their children's children for the sake of the
Divine Name.

BETWEEN ROSH HASHANAH AND YOM KIPPUR —
Remember us for life, O Sovereign who delights in life,
and inscribe us in the Book of Life, for Your sake, Living God.

Sovereign, Deliverer, Helper and Shield,
Blessed are You, Adonai, Sarah's Helper, Abraham's Shield.

בָּרוּךְ אַתָּה, יְיָ, מָגֵן אַבְרָהָם וְעֶזְרַת שָׂרָה.

אָבוֹת וְאִמָּהוֹת

גְּבוּרוֹת

קְדֻשָּׁה

בִּינָה

תְּשׁוּבָה

סְלִיחָה

גְּאֻלָּה

רְפוּאָה

בִּרְכַּת הַשָּׁנִים

חֵרוּת

מִשְׁפָּט

עַל הָרְשָׁעָה

צַדִּיקִים

יְרוּשָׁלַיִם

יְשׁוּעָה

שׁוֹמֵעַ תְּפִלָּה

עֲבוֹדָה

הוֹדָאָה

שָׁלוֹם

תְּפִלַּת הַלֵּב

For those who choose: At the beginning and end of the blessing, one bends the knees and bows
from the waist at the word בָּרוּךְ *Baruch* and stands straight at the word יְיָ *Adonai*.

I In the *Avot v'Imahot* we remind God of who we are by claiming illustrious ancestry. What qualities of which ancestors would you want to claim on your behalf?

My mom has been great. She has been doing nice things for me She has been letting me go outside at night and look for deer in my backyard.

?

The עֲמִידָה *Amidah* is recited while standing in reverent attention facing the ark. The *Amidah* is also known as the שְׁמוֹנֶה עֶשְׂרֵה *Sh'moneh Esreih* (Eighteen Blessings) and the תְּפִלָּה *T'filah* (Prayer), reflecting its significance as the core unit of the liturgy. On weekdays six blessings of praise and thirteen request-blessings (petitions) comprise the *Amidah*. On Shabbat and Festivals the *Amidah* contains only the seven blessings of praise.

אָבוֹת וְאִמָּהוֹת
גְּבוּרוֹת
קְדֻשָּׁה
בִּינָה
תְּשׁוּבָה
סְלִיחָה
גְּאֻלָּה
רְפוּאָה
בִּרְכַּת הַשָּׁנִים
חֵרוּת
מִשְׁפָּט
עַל הָרִשְׁעָה
צַדִּיקִים
יְרוּשָׁלַיִם
יְשׁוּעָה
שׁוֹמֵעַ תְּפִלָּה
עֲבוֹדָה
הוֹדָאָה
שָׁלוֹם
תְּפִלַּת הַלֵּב

אַתָּה גִּבּוֹר לְעוֹלָם, אֲדֹנָי,
מְחַיֵּה הַכֹּל (מֵתִים) אַתָּה,
רַב לְהוֹשִׁיעַ.

*WINTER — מַשִּׁיב הָרְוּחַ
וּמוֹרִיד הַגָּֽשֶׁם.
*SUMMER — מוֹרִיד הַטָּל.

מְכַלְכֵּל חַיִּים בְּחֶֽסֶד,
מְחַיֵּה הַכֹּל (מֵתִים)
בְּרַחֲמִים רַבִּים, סוֹמֵךְ נוֹפְלִים,
וְרוֹפֵא חוֹלִים, וּמַתִּיר אֲסוּרִים,
וּמְקַיֵּם אֱמוּנָתוֹ לִישֵׁנֵי עָפָר.
מִי כָמֽוֹךָ בַּֽעַל גְּבוּרוֹת
וּמִי דּֽוֹמֶה לָּךְ, מֶֽלֶךְ מֵמִית
וּמְחַיֶּה וּמַצְמִֽיחַ יְשׁוּעָה.

BETWEEN ROSH HASHANAH AND YOM KIPPUR — מִי כָמֽוֹךָ אַב הָרַחֲמִים,
זוֹכֵר יְצוּרָיו לְחַיִּים בְּרַחֲמִים.

וְנֶאֱמָן אַתָּה לְהַחֲיוֹת הַכֹּל (מֵתִים).
בָּרוּךְ אַתָּה, יְיָ, מְחַיֵּה הַכֹּל (הַמֵּתִים).

YOU ARE FOREVER MIGHTY, Adonai; You give life to all (revive the dead).

*WINTER — You cause the wind to shift and rain to fall.
*SUMMER — You rain dew upon us.

You sustain life through love, giving life to all (reviving the dead) through great compassion, supporting the fallen, healing the sick, freeing the captive, keeping faith with those who sleep in the dust. Who is like You, Source of mighty acts? Who resembles You, a Sovereign who takes and gives life, causing deliverance to spring up and faithfully giving life to all (reviving that which is dead)?

BETWEEN ROSH HASHANAH AND YOM KIPPUR — Who is like You, Compassionate God, who mercifully remembers Your creatures for life?

Blessed are You, Adonai, who gives life to all (revives the dead).

בָּרוּךְ אַתָּה, יְיָ, מְחַיֵּה הַכֹּל (הַמֵּתִים).

For Morning or Afternoon K'dushah, turn to pages 82–83.

אַתָּה גִּבּוֹר לְעוֹלָם *Atah gibor l'olam . . . You are forever mighty . . .* Is there nothing beyond God's ability? Historically, the *G'vurot* confronts the mystery of death in the face of God's power. God can reverse death. So it concludes, בָּרוּךְ אַתָּה, יְיָ, מְחַיֵּה הַמֵּתִים *Baruch atah, Adonai, m'chayeih hameitim, Blessed are You, Adonai, who revives the dead.* Our Reform tradition emphasizes life, and God's power to direct it in any way. בָּרוּךְ אַתָּה, יְיָ, מְחַיֵּה הַכֹּל *Baruch atah, Adonai, m'chayei hakol, Blessed are You, Adonai, who gives life to all.*

*WINTER: *Sh'mini Atzeret / Simchat Torah to Pesach* SUMMER: *Pesach to Sh'mini Atzeret / Simchat Torah*

E A good part of life is spent dealing with those who have greater power than we do — parents, teachers, bosses, those in authority. This prayer models one way to deal with the ultimate power. How does this serve as an example for dealing with human power?

מַשִׁיב הָרוּחַ / מוֹרִיד הַטָּל *Mashiv haruach / Morid hatal* — with these words, we join our Israeli brothers and sisters in their prayers for seasonal rains in the Land of Israel.

K'DUSHAH FOR WEEKDAY EVENING

אַתָּה קָדוֹשׁ וְשִׁמְךָ קָדוֹשׁ
וּקְדוֹשִׁים בְּכָל יוֹם
יְהַלְלוּךָ סֶּלָה.*
בָּרוּךְ אַתָּה, יְיָ, הָאֵל הַקָּדוֹשׁ.

*Between Rosh HaShanah and Yom Kippur —
בָּרוּךְ אַתָּה, יְיָ, הַמֶּלֶךְ הַקָּדוֹשׁ.

You are holy, Your Name is holy,
and those who are holy praise You every day.*
Blessed are You, Adonai, the holy God.
בָּרוּךְ אַתָּה, יְיָ, הָאֵל הַקָּדוֹשׁ.

*Between Rosh HaShanah and Yom Kippur —
Praised are You, Adonai, Holy Sovereign.
בָּרוּךְ אַתָּה, יְיָ, הַמֶּלֶךְ הַקָּדוֹשׁ.

Continue on pages 84–85.

This third blessing of praise in the עֲמִידָה *Amidah* emphasizes God's holy nature. Even God's Name is holy.

Avot v'Imahot

G'vurot

K'dushah

Binah

T'shuvah

S'lichah

G'ulah

R'fuah

Birkat HaShanim

Cheirut

Mishpat

Al HaRishah

Tzadikim

Y'rushalayim

Y'shuah

Shomei-a T'filah

Avodah

Hodaah

Shalom

T'filat HaLev

K'DUSHAH FOR WEEKDAY MORNING OR AFTERNOON

נְקַדֵּשׁ אֶת שִׁמְךָ בָּעוֹלָם,

כְּשֵׁם שֶׁמַּקְדִּישִׁים אוֹתוֹ בִּשְׁמֵי מָרוֹם,

כַּכָּתוּב עַל יַד נְבִיאֶךָ,

וְקָרָא זֶה אֶל זֶה וְאָמַר:

קָדוֹשׁ, קָדוֹשׁ, קָדוֹשׁ יְיָ צְבָאוֹת,

מְלֹא כָל הָאָרֶץ כְּבוֹדוֹ.

לְעֻמָּתָם בָּרוּךְ יֹאמֵרוּ:

בָּרוּךְ כְּבוֹד יְיָ מִמְּקוֹמוֹ.

וּבְדִבְרֵי קָדְשְׁךָ כָּתוּב לֵאמֹר:

יִמְלֹךְ יְיָ לְעוֹלָם, אֱלֹהַיִךְ צִיּוֹן

לְדֹר וָדֹר, הַלְלוּיָהּ.

LET US SANCTIFY Your name on earth as it is sanctified in the heavens above.
As written by Your prophet:
Holy, holy, holy is *Adonai Tz'vaot*! God's Presence fills all the earth.
They responded in blessing:
Blessed is the presence of God, shining forth from where God dwells.
In Your holy scripture it is written:
Adonai shall reign forever, Your God O Zion, for all generations, Hallelujah.

לְדוֹר וָדוֹר נַגִּיד גָּדְלֶךָ

וּלְנֵצַח נְצָחִים קְדֻשָּׁתְךָ נַקְדִּישׁ,

וְשִׁבְחֲךָ, אֱלֹהֵינוּ,

מִפִּינוּ לֹא יָמוּשׁ לְעוֹלָם וָעֶד.*

בָּרוּךְ אַתָּה, יְיָ, הָאֵל הַקָּדוֹשׁ.

FOR ALL GENERATIONS we will tell of Your greatness and for all eternity
proclaim Your holiness. Your praise, our God, will never depart from our mouths,
for You are a Sovereign God, great and holy.*
Blessed are You Adonai, the holy God.

בָּרוּךְ אַתָּה, יְיָ, הָאֵל הַקָּדוֹשׁ.

*BETWEEN ROSH HASHANAH AND YOM KIPPUR —
Blessed are You, Adonai, Holy Sovereign.

בָּרוּךְ אַתָּה, יְיָ, הַמֶּלֶךְ הַקָּדוֹשׁ.

For those who choose: At the words וְקָרָא זֶה *v'kara zeh* one bows to the left and at אֶל זֶה *el zeh*
one bows to the right, and at each mention of קָדוֹשׁ *kadosh*, one rises on one's toes.

יְיָ צְבָאוֹת *Adonai Tz'vaot* . . . God is portrayed as having a heavenly array.

קָדוֹשׁ, קָדוֹשׁ, קָדוֹשׁ *Kadosh, kadosh, kadosh* . . . *Holy, holy, holy* . . . Isaiah 6:3

בָּרוּךְ כְּבוֹד *Baruch k'vod* . . . *Blessed is the presence* . . . Ezekiel 3:12

יִמְלֹךְ יְיָ לְעוֹלָם *Yimloch Adonai l'olam* . . . *Adonai shall reign forever* . . . Psalm 146:10

אָבוֹת וְאִמָּהוֹת
גְּבוּרוֹת
קְדֻשָּׁה

בִּינָה
תְּשׁוּבָה
סְלִיחָה
גְּאֻלָּה
רְפוּאָה
בִּרְכַּת הַשָּׁנִים
חֵרוּת
מִשְׁפָּט
עַל הָרִשְׁעָה
צַדִּיקִים
יְרוּשָׁלַיִם
יְשׁוּעָה
שׁוֹמֵעַ תְּפִלָּה

עֲבוֹדָה
הוֹדָאָה
שָׁלוֹם
תְּפִלַּת הַלֵּב

Avot v'Imahot

G'vurot

K'dushah

Binah

T'shuvah

S'lichah

G'ulah

R'fuah

Birkat HaShanim

Cheirut

Mishpat

Al HaRishah

Tzadikim

Y'rushalayim

Y'shuah

Shomei-a T'filah

Avodah

Hodaah

Shalom

T'filat HaLev

אָבוֹת וְאִמָּהוֹת
גְּבוּרוֹת
קְדֻשָּׁה
בִּינָה
תְּשׁוּבָה
סְלִיחָה
גְּאֻלָּה
רְפוּאָה
בִּרְכַּת הַשָּׁנִים
חֵרוּת
מִשְׁפָּט
עַל הָרִשְׁעָה
צַדִּיקִים
יְרוּשָׁלַיִם
יְשׁוּעָה
שׁוֹמֵעַ תְּפִלָּה
עֲבוֹדָה
הוֹדָאָה
שָׁלוֹם
תְּפִלַּת הַלֵּב

אַתָּה חוֹנֵן לְאָדָם דַּעַת
וּמְלַמֵּד לֶאֱנוֹשׁ בִּינָה.
חָנֵּנוּ מֵאִתְּךָ
חָכְמָה, בִּינָה וְדָעַת.
בָּרוּךְ אַתָּה, יְיָ, חוֹנֵן הַדָּעַת.

YOU GRACE humans with knowledge
and teach mortals understanding.
Graciously share with us Your wisdom, insight, and knowledge.
Blessed are You, Adonai, who graces us with knowledge.

הֲשִׁיבֵנוּ לְתוֹרָתֶךָ
וְקָרְבֵנוּ לַעֲבוֹדָתֶךָ
וְהַחֲזִירֵנוּ בִּתְשׁוּבָה שְׁלֵמָה לְפָנֶיךָ.
בָּרוּךְ אַתָּה, יְיָ, הָרוֹצֶה בִּתְשׁוּבָה.

RETURN US to Your Torah and draw us to Your service,
and in complete repentance restore us to Your Presence.
Blessed are You, Adonai, who welcomes repentance.

סְלַח לָנוּ כִּי חָטָאנוּ,
מְחַל לָנוּ כִּי פָשָׁעְנוּ,
כִּי מוֹחֵל וְסוֹלֵחַ אָתָּה.
בָּרוּךְ אַתָּה, יְיָ, הַמַּרְבֶּה לִסְלֹחַ.

FORGIVE US for we have sinned, pardon us for we have transgressed,
for You pardon and forgive.
Blessed are You, Adonai, abounding in forgiveness.

רְאֵה בְעָנְיֵנוּ וְרִיבָה רִיבֵנוּ,
וּגְאָלֵנוּ מְהֵרָה לְמַעַן שְׁמֶךָ,
כִּי גוֹאֵל חָזָק אָתָּה.
בָּרוּךְ אַתָּה, יְיָ, גּוֹאֵל יִשְׂרָאֵל.

TAKE NOTE OF our affliction and make our struggles Yours.
Redeem us swiftly for Your Name's sake,
for You are the mighty redeemer.
Blessed are You, Adonai, who redeems Israel.

The middle blessings of the עֲמִידָה *Amidah* are petitions to God; they are offered only on weekdays because Shabbat is a time of perfect peace, making petition unnecessary.

Avot v'Imahot

G'vurot

K'dushah

Binah

T'shuvah

S'lichah

G'ulah

R'fuah

Birkat HaShanim

Cheirut

Mishpat

Al HaRishah

Tzadikim

Y'rushalayim

Y'shuah

Shomei-a T'filah

Avodah

Hodaah

Shalom

T'filat HaLev

רְפָאֵנוּ, יְיָ, וְנֵרָפֵא,
הוֹשִׁיעֵנוּ וְנִוָּשֵׁעָה,
וְהַעֲלֵה רְפוּאָה שְׁלֵמָה
לְכָל מַכּוֹתֵינוּ
וּלְכָל תַּחֲלוּאֵינוּ
וּלְכָל מַכְאוֹבֵינוּ.
בָּרוּךְ אַתָּה, יְיָ, רוֹפֵא הַחוֹלִים.

HEAL US, Adonai, and let us be healed; save us and let us be saved.
Grant full healing to our every illness, wound and pain.
Blessed are You, Adonai, who heals the sick.

בָּרֵךְ עָלֵינוּ אֶת הַשָּׁנָה הַזֹּאת
וְאֶת כָּל מִינֵי תְבוּאָתָהּ לְטוֹבָה.
וְתֵן בְּרָכָה עַל פְּנֵי הָאֲדָמָה,
וְשַׂבְּעֵנוּ מִטּוּבֶךָ.
בָּרוּךְ אַתָּה, יְיָ, מְבָרֵךְ הַשָּׁנִים.

BLESS this our year and its abundant harvest for good.
Grant blessing throughout the earth
and satisfy us with Your goodness.
Blessed are You, Adonai, who blesses the years.

תְּקַע בְּשׁוֹפָר גָּדוֹל לְחֵרוּתֵנוּ,
וְשָׂא נֵס גָּדוֹל לַעֲשׁוּקֵינוּ,
וְקוֹל דְּרוֹר יִשָּׁמַע
בְּאַרְבַּע כַּנְפוֹת הָאָרֶץ.
בָּרוּךְ אַתָּה, יְיָ, פּוֹדֶה עֲשׁוּקִים.

SOUND the great shofar to proclaim our freedom,
raise a great banner for our oppressed
and let the voice of liberty be heard in the four corners of the earth.
Blessed are You, Adonai, who redeems the oppressed.

Avot v'Imahot

G'vurot

K'dushah

Binah

T'shuvah

S'lichah

G'ulah

R'fuah

Birkat HaShanim

Cheirut

Mishpat

Al HaRishah

Tzadikim

Y'rushalayim

Y'shuah

Shomei-a T'filah

Avodah

Hodaah

Shalom

T'filat HaLev

אָבוֹת וְאִמָּהוֹת
גְּבוּרוֹת
קְדֻשָּׁה
בִּינָה
תְּשׁוּבָה
סְלִיחָה
גְּאֻלָּה
רְפוּאָה
בִּרְכַּת הַשָּׁנִים
חֵרוּת
מִשְׁפָּט
עַל הָרִשְׁעָה
צַדִּיקִים
יְרוּשָׁלַיִם
יְשׁוּעָה
שׁוֹמֵעַ תְּפִלָּה
עֲבוֹדָה
הוֹדָאָה
שָׁלוֹם
תְּפִלַּת הַלֵּב

עַל שׁוֹפְטֵי אֶרֶץ שְׁפוֹךְ רוּחֶךָ
וְהַדְרִיכֵם בְּמִשְׁפְּטֵי צִדְקֶךָ,
וּמְלֹךְ עָלֵינוּ אַתָּה לְבַדְּךָ
בְּחֶסֶד וּבְרַחֲמִים.
בָּרוּךְ אַתָּה, יְיָ,
אוֹהֵב צְדָקָה וּמִשְׁפָּט.

POUR Your spirit upon the rulers of all lands;
guide them that they may govern justly.
O may You alone rule over us in steadfast love and compassion.
Blessed are You, Adonai, who loves righteousness and justice.

וְלָרִשְׁעָה אַל תְּהִי תִקְוָה,
וְהַתּוֹעִים אֵלֶיךָ יָשׁוּבוּ,
וּמַלְכוּת זָדוֹן מְהֵרָה תְּשַׁבֵּר.
בָּרוּךְ אַתָּה, יְיָ,
שׁוֹבֵר רֶשַׁע מִן הָאָרֶץ.

AND FOR WICKEDNESS, let there be no hope,
and may all the errant return to You,
and may the realm of wickedness be shattered.
Blessed are You, Adonai, whose will it is
that the wicked vanish from the earth.

עַל הַצַּדִּיקִים וְעַל הַחֲסִידִים
וְעַל זִקְנֵי עַמְּךָ בֵּית יִשְׂרָאֵל,
וְעַל גֵּרֵי הַצֶּדֶק וְעָלֵינוּ,
יֶהֱמוּ נָא רַחֲמֶיךָ, יְיָ אֱלֹהֵינוּ,
וְתֵן שָׂכָר טוֹב
לְכָל הַבּוֹטְחִים בְּשִׁמְךָ בֶּאֱמֶת,
וְשִׂים חֶלְקֵנוּ עִמָּהֶם לְעוֹלָם.
בָּרוּךְ אַתָּה, יְיָ,
מִשְׁעָן וּמִבְטָח לַצַּדִּיקִים.

TOWARD THE RIGHTEOUS, toward the pious,
toward the leaders of Your people Israel,
toward those who choose sincerely to be Jews and toward us all,
may Your tender mercies be stirred.
Adonai, our God, grant a good reward to all who trust in Your name
and number us among them.
Blessed is Adonai, the staff and the stay of the righteous.

Avot v'Imahot

G'vurot

K'dushah

Binah

T'shuvah

S'lichah

G'ulah

R'fuah

Birkat HaShanim

Cheirut

Mishpat

Al HaRishah

Tzadikim

Y'rushalayim

Y'shuah

Shomei-a T'filah

Avodah

Hodaah

Shalom

T'filat HaLev

אָבוֹת וְאִמָּהוֹת
גְּבוּרוֹת
קְדֻשָּׁה
בִּינָה
תְּשׁוּבָה
סְלִיחָה
גְּאֻלָּה
רְפוּאָה
בִּרְכַּת הַשָּׁנִים
חֵרוּת
מִשְׁפָּט
עַל הָרִשְׁעָה
צַדִּיקִים
יְרוּשָׁלַיִם
יְשׁוּעָה
שׁוֹמֵעַ תְּפִלָּה
עֲבוֹדָה
הוֹדָאָה
שָׁלוֹם
תְּפִלַּת הַלֵּב

וְלִירוּשָׁלַיִם
עִירְךָ בְּרַחֲמִים תִּפְנֶה,
וִיהִי שָׁלוֹם בִּשְׁעָרֶיהָ
וְשַׁלְוָה בְּלֵב יוֹשְׁבֶיהָ,
וְתוֹרָתְךָ מִצִּיּוֹן תֵּצֵא,
וּדְבָרְךָ מִירוּשָׁלָיִם.
בָּרוּךְ אַתָּה, יְיָ,
נוֹתֵן שָׁלוֹם בִּירוּשָׁלָיִם.

AND TURN in compassion to Jerusalem, Your city.
May there be peace in her gates, quietness in the hearts of her inhabitants.
Let Your Torah go forth from Zion and Your word from Jerusalem.
Blessed is Adonai, who gives peace to Jerusalem.

אֱמֶת מֵאֶרֶץ תִּצְמָח
וְצֶדֶק מִשָּׁמַיִם נִשְׁקָף,
וְקֶרֶן עַמְּךָ תָּרוּם בִּישׁוּעָתֶךָ,
כִּי לִישׁוּעָתְךָ קִוִּינוּ כָּל הַיּוֹם.
בָּרוּךְ אַתָּה, יְיָ, מַצְמִיחַ קֶרֶן יְשׁוּעָה.

MAY TRUTH spring up from the earth;
May justice look down from the heavens.
May the strength of Your people flourish through
Your deliverance for we continually hope for Your deliverance.
Blessed are You, Adonai, who causes salvation to flourish.

שְׁמַע קוֹלֵנוּ, יְיָ אֱלֹהֵינוּ,
חוּס וְרַחֵם עָלֵינוּ,
וְקַבֵּל בְּרַחֲמִים וּבְרָצוֹן אֶת תְּפִלָּתֵנוּ,
כִּי אֵל שׁוֹמֵעַ תְּפִלּוֹת וְתַחֲנוּנִים אָתָּה.
בָּרוּךְ אַתָּה, יְיָ, שׁוֹמֵעַ תְּפִלָּה.

HEAR our voice, Adonai our God,
have compassion upon us and accept our prayer with favor and mercy,
for You are a God who hears prayer and supplication.
Blessed is Adonai, who hearkens to prayer.

אֱמֶת מֵאֶרֶץ תִּצְמָח *Emet mei-eretz titzmach . . . May truth spring up from the earth . . .* Psalm 85:12

Avot v'Imahot
G'vurot
K'dushah

Binah
T'shuvah
S'lichah
G'ulah
R'fuah
Birkat HaShanim
Cheirut
Mishpat
Al HaRishah
Tzadikim
Y'rushalayim
Y'shuah
Shomei-a T'filah

Avodah
Hodaah
Shalom
T'filat HaLev

רְצֵה, יְיָ אֱלֹהֵינוּ, בְּעַמְּךָ יִשְׂרָאֵל,
וּתְפִלָּתָם בְּאַהֲבָה תְקַבֵּל,
וּתְהִי לְרָצוֹן תָּמִיד
עֲבוֹדַת יִשְׂרָאֵל עַמֶּךָ.
אֵל קָרוֹב לְכָל קֹרְאָיו,
פְּנֵה אֶל עֲבָדֶיךָ וְחָנֵּנוּ,
שְׁפוֹךְ רוּחֲךָ עָלֵינוּ.

FIND FAVOR, Adonai, our God, with Your people Israel
and accept their prayer in love.
May the worship of Your people Israel always be acceptable.
God who is near to all who call, turn lovingly to Your servants.
Pour out Your spirit upon us.

ROSH CHODESH, PESACH, AND SUKKOT

אֱלֹהֵינוּ וֵאלֹהֵי אֲבוֹתֵינוּ וְאִמּוֹתֵינוּ,
יַעֲלֶה וְיָבֹא וְיִזָּכֵר זִכְרוֹנֵנוּ
וְזִכְרוֹן כָּל עַמְּךָ בֵּית יִשְׂרָאֵל לְפָנֶיךָ,
לְטוֹבָה, לְחֵן וּלְחֶסֶד וּלְרַחֲמִים,
לְחַיִּים וּלְשָׁלוֹם, בְּיוֹם

רֹאשׁ הַחֹדֶשׁ הַזֶּה.
חַג הַמַּצוֹת הַזֶּה.
חַג הַסֻּכּוֹת הַזֶּה.
זָכְרֵנוּ, יְיָ אֱלֹהֵינוּ, בּוֹ לְטוֹבָה. אָמֵן.
וּפָקְדֵנוּ בוֹ לִבְרָכָה. אָמֵן.
וְהוֹשִׁיעֵנוּ בוֹ לְחַיִּים. אָמֵן.

Our God and God of our fathers and mothers, on this
(first day of the new month) — (day of Pesach) — (day of Sukkot)
be mindful of us and all Your people Israel,
for good, for love, for compassion, life and peace.
Remember us for wellbeing. Amen.
Visit us with blessing. Amen. Help us to a fuller life. Amen.

וְתֶחֱזֶינָה עֵינֵינוּ בְּשׁוּבְךָ
לְצִיּוֹן בְּרַחֲמִים.

LET OUR EYES BEHOLD Your loving return to Zion.
Blessed are You, Adonai, whose Presence returns to Zion.

בָּרוּךְ אַתָּה, יְיָ,
הַמַּחֲזִיר שְׁכִינָתוֹ לְצִיּוֹן.

אָבוֹת וְאִמָּהוֹת

גְּבוּרוֹת

קְדֻשָּׁה

בִּינָה

תְּשׁוּבָה

סְלִיחָה

גְּאֻלָּה

רְפוּאָה

בִּרְכַּת הַשָּׁנִים

חֵרוּת

מִשְׁפָּט

עַל הָרְשָׁעָה

צַדִּיקִים

יְרוּשָׁלַיִם

יְשׁוּעָה

שׁוֹמֵעַ תְּפִלָּה

עֲבוֹדָה

הוֹדָאָה

שָׁלוֹם

תְּפִלַּת הַלֵּב

T The prayerbook used in Israeli Reform synagogues is called הָעֲבוֹדָה שֶׁבַּלֵּב *HaAvodah Shebalev,* "Service [or Prayer] of the Heart." What do you think this phrase means? Why do you think it is used for the title of a prayerbook?

אָבוֹת וְאִמָּהוֹת

גְּבוּרוֹת

קְדֻשָּׁה

בִּינָה

תְּשׁוּבָה

סְלִיחָה

גְּאֻלָּה

רְפוּאָה

בִּרְכַּת הַשָּׁנִים

חֵרוּת

מִשְׁפָּט

עַל הָרְשָׁעָה

צַדִּיקִים

יְרוּשָׁלַיִם

יְשׁוּעָה

שׁוֹמֵעַ תְּפִלָּה

עֲבוֹדָה

הוֹדָאָה

שָׁלוֹם

תְּפִלַּת הַלֵּב

מוֹדִים אֲנַחְנוּ לָךְ, שָׁאַתָּה הוּא
יְיָ אֱלֹהֵינוּ וֵאלֹהֵי אֲבוֹתֵינוּ וְאִמּוֹתֵינוּ
לְעוֹלָם וָעֶד. צוּר חַיֵּינוּ, מָגֵן יִשְׁעֵנוּ,
אַתָּה הוּא לְדוֹר וָדוֹר.

נוֹדֶה לְּךָ וּנְסַפֵּר תְּהִלָּתֶךָ. עַל חַיֵּינוּ
הַמְּסוּרִים בְּיָדֶךָ, וְעַל נִשְׁמוֹתֵינוּ
הַפְּקוּדוֹת לָךְ, וְעַל נִסֶּיךָ שֶׁבְּכָל יוֹם
עִמָּנוּ, וְעַל נִפְלְאוֹתֶיךָ וְטוֹבוֹתֶיךָ
שֶׁבְּכָל עֵת, עֶרֶב וָבֹקֶר וְצָהֳרָיִם.

הַטּוֹב כִּי לֹא כָלוּ רַחֲמֶיךָ, וְהַמְרַחֵם
כִּי לֹא תַמּוּ חֲסָדֶיךָ, מֵעוֹלָם קִוִּינוּ לָךְ.

WE ACKNOWLEDGE with thanks that You are Adonai, our God and the God of
our ancestors, forever. You are the Rock of our lives, and the Shield of our salvation in
every generation. Let us thank You and praise You — for our lives which are in Your
hand, for our souls which are in Your care, for Your miracles that we experience every
day and for Your wondrous deeds and favors at every time of day: evening, morning
and noon. O Good One, whose mercies never end, O Compassionate One, whose
kindness never fails, we forever put our hope in You.

וְעַל כֻּלָּם יִתְבָּרַךְ וְיִתְרוֹמַם שִׁמְךָ,
מַלְכֵּנוּ, תָּמִיד לְעוֹלָם וָעֶד.

BETWEEN ROSH HASHANAH AND
YOM KIPPUR — וּכְתוֹב לְחַיִּים
טוֹבִים כָּל בְּנֵי בְרִיתֶךָ.

וְכֹל הַחַיִּים יוֹדוּךָ סֶּלָה, וִיהַלְלוּ
אֶת שִׁמְךָ בֶּאֱמֶת, הָאֵל יְשׁוּעָתֵנוּ
וְעֶזְרָתֵנוּ סֶלָה. בָּרוּךְ אַתָּה, יְיָ,
הַטּוֹב שִׁמְךָ וּלְךָ נָאֶה לְהוֹדוֹת.

For all these things, O Sovereign, let Your Name be forever praised and blessed.
BETWEEN ROSH HASHANAH AND YOM KIPPUR — Inscribe all the children
of Your covenant for a good life.

O God, our Redeemer and Helper, let all who live affirm You and praise Your Name in
truth. Blessed are You, Adonai, Your Name is Goodness, and You are worthy of thanksgiving.

בָּרוּךְ אַתָּה, יְיָ, הַטּוֹב שִׁמְךָ וּלְךָ נָאֶה לְהוֹדוֹת.

I Wrap a present for someone you care about very much. Carefully fold the wrapping paper, and decorate the gift that you are giving with your own special touch. Imagine the look in the eyes of that special person as he or she opens your gift. Now, turn that imagery around and imagine that you were the recipient of such a gift. Describe your gratitude.

For those who choose: At the word מוֹדִים *modim,* one bows at the waist. At יְיָ *Adonai,* one stands up straight.

BIRKAT SHALOM FOR WEEKDAY AFTERNOON AND EVENING

אָבוֹת וְאִמָּהוֹת
גְּבוּרוֹת
קְדֻשָּׁה
בִּינָה
תְּשׁוּבָה
סְלִיחָה
גְּאֻלָּה
רְפוּאָה
בִּרְכַּת הַשָּׁנִים
חֵרוּת
מִשְׁפָּט
עַל הָרִשְׁעָה
צַדִּיקִים
יְרוּשָׁלַיִם
יְשׁוּעָה
שׁוֹמֵעַ תְּפִלָּה
עֲבוֹדָה
הוֹדָאָה
שָׁלוֹם
תְּפִלַּת הַלֵּב

שָׁלוֹם רָב עַל יִשְׂרָאֵל עַמְּךָ
תָּשִׂים לְעוֹלָם,
כִּי אַתָּה הוּא מֶלֶךְ אָדוֹן
לְכָל הַשָּׁלוֹם.
וְטוֹב בְּעֵינֶיךָ לְבָרֵךְ
אֶת עַמְּךָ יִשְׂרָאֵל
בְּכָל עֵת וּבְכָל שָׁעָה בִּשְׁלוֹמֶךָ.

BETWEEN ROSH HASHANAH AND YOM KIPPUR — בְּסֵפֶר חַיִּים,
בְּרָכָה וְשָׁלוֹם
וּפַרְנָסָה טוֹבָה,
נִזָּכֵר וְנִכָּתֵב לְפָנֶיךָ,
אֲנַחְנוּ וְכָל עַמְּךָ בֵּית יִשְׂרָאֵל,
לְחַיִּים טוֹבִים וּלְשָׁלוֹם.
בָּרוּךְ אַתָּה, יְיָ, עוֹשֵׂה הַשָּׁלוֹם.

בָּרוּךְ אַתָּה, יְיָ,
הַמְבָרֵךְ אֶת עַמּוֹ יִשְׂרָאֵל בַּשָּׁלוֹם.

GRANT ABUNDANT peace to Israel Your people forever,
for You are the sovereign God of all peace.
May it be pleasing to You to bless Your people Israel,
in every season and moment with Your peace.

BETWEEN ROSH HASHANAH AND YOM KIPPUR — In the book of life, blessing, peace and prosperity, may we be remembered and inscribed by You, we and all Your people Israel for a good life and for peace. Blessed are You, Adonai, who makes peace.

בָּרוּךְ אַתָּה, יְיָ, עוֹשֵׂה הַשָּׁלוֹם.

Blessed are You, Adonai, who blesses Your people Israel with peace.

בָּרוּךְ אַתָּה, יְיָ, הַמְבָרֵךְ אֶת עַמּוֹ יִשְׂרָאֵל בַּשָּׁלוֹם.

Continue on pages 100–101.

Avot v'Imahot
G'vurot
K'dushah

Binah
T'shuvah
S'lichah
G'ulah
R'fuah
Birkat HaShanim
Cheirut
Mishpat
Al HaRishah
Tzadikim
Y'rushalayim
Y'shuah
Shomei-a T'filah

Avodah
Hodaah
Shalom
T'filat HaLev

P שָׁלוֹם רָב, *Shalom rav,* can be understood as "an abundance of peace." Why is it that a little amount of peace is not enough for us? What do we stand to gain from "abundant peace"?

BIRKAT SHALOM FOR WEEKDAY MORNING

שִׂים שָׁלוֹם טוֹבָה וּבְרָכָה,

חֵן וָחֶסֶד וְרַחֲמִים,

עָלֵינוּ וְעַל כָּל יִשְׂרָאֵל עַמֶּךָ.

בָּרְכֵנוּ, יוֹצְרֵנוּ, כֻּלָּנוּ כְּאֶחָד

בְּאוֹר פָּנֶיךָ, כִּי בְאוֹר פָּנֶיךָ נָתַתָּ לָּנוּ,

יְיָ אֱלֹהֵינוּ,

תּוֹרַת חַיִּים וְאַהֲבַת חֶסֶד,

וּצְדָקָה וּבְרָכָה וְרַחֲמִים

וְחַיִּים וְשָׁלוֹם.

וְטוֹב בְּעֵינֶיךָ לְבָרֵךְ

אֶת עַמְּךָ יִשְׂרָאֵל

בְּכָל עֵת וּבְכָל שָׁעָה בִּשְׁלוֹמֶךָ.

BETWEEN ROSH HASHANAH AND

בְּסֵפֶר חַיִּים, — YOM KIPPUR

בְּרָכָה וְשָׁלוֹם, וּפַרְנָסָה טוֹבָה

נִזָּכֵר וְנִכָּתֵב לְפָנֶיךָ,

אֲנַחְנוּ וְכָל עַמְּךָ בֵּית יִשְׂרָאֵל,

לְחַיִּים טוֹבִים וּלְשָׁלוֹם.

בָּרוּךְ אַתָּה, יְיָ,

הַמְבָרֵךְ אֶת עַמּוֹ יִשְׂרָאֵל בַּשָּׁלוֹם.

אָבוֹת וְאִמָּהוֹת

גְּבוּרוֹת

קְדֻשָּׁה

בִּינָה

תְּשׁוּבָה

סְלִיחָה

גְּאֻלָּה

רְפוּאָה

בִּרְכַּת הַשָּׁנִים

חֵרוּת

מִשְׁפָּט

עַל הָרְשָׁעָה

צַדִּיקִים

יְרוּשָׁלַיִם

יְשׁוּעָה

שׁוֹמֵעַ תְּפִלָּה

עֲבוֹדָה

הוֹדָאָה

שָׁלוֹם

תְּפִלַּת הַלֵּב

GRANT PEACE, goodness and blessing, grace, kindness, and mercy,
to us and to all Your people Israel.
Bless us, our Creator, all of us together, through the light of Your Presence.
Truly through the light of Your Presence, Adonai our God, You gave us
a Torah of life — the love of kindness, justice and blessing, mercy, life, and peace.
May You see fit to bless Your people Israel, at all times, at every hour,
with Your peace.

BETWEEN ROSH HASHANAH AND YOM KIPPUR — Inscribe us for life, blessing, peace and
prosperity, remembering all Your people Israel for life and peace. Blessed are You, Adonai,
Source of peace.

בָּרוּךְ אַתָּה, יְיָ, עוֹשֶׂה הַשָּׁלוֹם.

Praised are You Adonai, who blesses Your people Israel with peace.

בָּרוּךְ אַתָּה, יְיָ, הַמְבָרֵךְ אֶת עַמּוֹ יִשְׂרָאֵל בַּשָּׁלוֹם.

P This prayer asks for God to grant peace to God's people Israel. Some prayerbooks change this prayer to ask for peace for all God's people, for all the world, or for all who dwell in nature. Do you think that it is appropriate to ask for peace for ourselves, or do you think we should pray for peace for all?

אֱלֹהַי, נְצֹר לְשׁוֹנִי מֵרָע וּשְׂפָתַי
מִדַּבֵּר מִרְמָה, וְלִמְקַלְלַי נַפְשִׁי
תִדֹּם, וְנַפְשִׁי כֶּעָפָר לַכֹּל תִּהְיֶה.
פְּתַח לִבִּי בְּתוֹרָתֶךָ, וּבְמִצְוֹתֶיךָ
תִּרְדֹּף נַפְשִׁי. וְכֹל הַחוֹשְׁבִים
עָלַי רָעָה, מְהֵרָה הָפֵר עֲצָתָם
וְקַלְקֵל מַחֲשַׁבְתָּם. עֲשֵׂה לְמַעַן
שְׁמֶךָ, עֲשֵׂה לְמַעַן יְמִינֶךָ, עֲשֵׂה
לְמַעַן קְדֻשָּׁתֶךָ, עֲשֵׂה לְמַעַן
תּוֹרָתֶךָ. לְמַעַן יֵחָלְצוּן יְדִידֶיךָ,
הוֹשִׁיעָה יְמִינְךָ וַעֲנֵנִי.

My God, guard my speech from evil and my lips from deception.
Before those who slander me, I will hold my tongue; I will practice humility.
Open my heart to Your Torah, that I may pursue Your mitzvot.
As for all who think evil of me, cancel their designs and frustrate their schemes.
Act for your own sake, for the sake of Your Power,
for the sake of Your Holiness, for the sake of Your Torah,
so that Your loved ones may be rescued. Save with Your power,
and answer me.

יִהְיוּ לְרָצוֹן אִמְרֵי פִי וְהֶגְיוֹן לִבִּי
לְפָנֶיךָ, יְיָ צוּרִי וְגוֹאֲלִי.

May the words of my mouth and the meditations of my heart
be acceptable to You, Adonai, my Rock and my Redeemer.

עֹשֶׂה שָׁלוֹם בִּמְרוֹמָיו, הוּא יַעֲשֶׂה
שָׁלוֹם עָלֵינוּ, וְעַל כָּל יִשְׂרָאֵל
וְעַל כָּל יוֹשְׁבֵי תֵבֵל, וְאִמְרוּ: אָמֵן.

May the One who makes peace in the high heavens
make peace for us, all Israel and all who inhabit the earth. Amen.

Reading of the Torah is on page 104. Aleinu and Kaddish begin on page 586.

אָבוֹת וְאִמָּהוֹת
גְּבוּרוֹת
קְדֻשָּׁה
בִּינָה
תְּשׁוּבָה
סְלִיחָה
גְּאֻלָּה
רְפוּאָה
בִּרְכַּת הַשָּׁנִים
חֵרוּת
מִשְׁפָּט
עַל הָרִשְׁעָה
צַדִּיקִים
יְרוּשָׁלַיִם
יְשׁוּעָה
שׁוֹמֵעַ תְּפִלָּה
עֲבוֹדָה
הוֹדָאָה
שָׁלוֹם
תְּפִלַּת הַלֵּב

אֱלֹהַי, נְצֹר *Elohai, n'tzor . . . My God, guard . . . based on* Psalm 34:14
לְמַעַן יֵחָלְצוּן *L'maan yeichaltzun . . . so that Your loved ones . . .* Psalm 60:7
יִהְיוּ לְרָצוֹן *Yih'yu l'ratzon . . . May the words of my mouth . . .* Psalm 19:15

Avot v'Imahot
G'vurot
K'dushah

Binah
T'shuvah
S'lichah
G'ulah
R'fuah
Birkat HaShanim
Cheirut
Mishpat
Al HaRishah
Tzadikim
Y'rushalayim
Y'shuah
Shomei-a T'filah

Avodah
Hodaah
Shalom
T'filat HaLev

P Let go of all written prayers. Try to pray directly to God, without the intermediary of language. Is this satisfying, or do you need written prayers in order to have a meaningful prayer experience?

מֵעֵין שְׁמוֹנֶה עֶשְׂרֵה

אָנָּא, יְיָ, חָנֵּנוּ מֵאִתְּךָ דֵּעָה.
אָנָּא, חוֹנֵן הַדָּעַת,
הַחֲזִירֵנוּ בִּתְשׁוּבָה שְׁלֵמָה לְפָנֶיךָ.
אָנָּא, הָרוֹצֶה בִּתְשׁוּבָה,
סְלַח לָנוּ כִּי חָטָאנוּ.
אָנָּא, חַנּוּן הַמַּרְבֶּה לִסְלוֹחַ,
גְּאָלֵנוּ לְמַעַן שְׁמֶךָ.
אָנָּא, גּוֹאֵל יִשְׂרָאֵל, רְפָאֵנוּ וְנֵרָפֵא.
אָנָּא, רוֹפֵא הַחוֹלִים,
בָּרֵךְ שְׁנוֹתֵינוּ מִטּוּבֶךָ.
אָנָּא, מְבָרֵךְ הַשָּׁנִים,
שָׂא נֵס לִפְדּוֹת עֲשׁוּקֵינוּ.
אָנָּא, פּוֹדֶה עֲשׁוּקִים,
מְלֹךְ עָלֵינוּ בְּמִשְׁפְּטֵי צִדְקֶךָ.
אָנָּא, מֶלֶךְ אוֹהֵב צְדָקָה וּמִשְׁפָּט,
יֶהֱמוּ רַחֲמֶיךָ עַל צַדִּיקֶיךָ.
אָנָּא, מִשְׁעָן וּמִבְטָח לַצַּדִּיקִים,
תֵּן שָׁלוֹם בִּירוּשָׁלָיִם.
אָנָּא, יְיָ, שְׁמַע קוֹלֵנוּ,
קַבֵּל תַּחֲנוּנֵינוּ, וְשִׂים שָׁלוֹם בָּעוֹלָם.

ADONAI, OUR GOD grant us knowledge.
Having granted us knowledge, accept our repentance.
Having accepted our repentance, forgive our sins.
Having forgiven us, redeem us.
Having redeemed us, heal our ills.
Having healed us, bless our years.
Having blessed our years, gather us together.
Having gathered us together, judge us righteously.
Having judged us righteously, defeat evil.
Having defeated evil, let goodness flourish.
Now that the righteous flourish, make Jerusalem heavenly for us.
Hear our entreaties and accept us. In Your great goodness, grant Jerusalem peace.

בָּרוּךְ אַתָּה, יְיָ, שׁוֹמֵעַ תְּפִלָּה.
Blessed are You, Adonai, who hears prayer.

Mei-ein Sh'moneh Esrei

מֵעֵין שְׁמוֹנֶה עֶשְׂרֵה *Mei-ein Sh'moneh Esrei* is a short form of the עֲמִידָה *Amidah,* various versions of which were used by the Rabbis of antiquity. We contribute our own here with language drawn from tradition, but expressive of contemporary yearning.

קַבָּלַת הַתּוֹרָה

הַקָּפָה

בִּרְכוֹת הַתּוֹרָה

מִי שֶׁבֵּרַךְ

הַגְבָּהָה

בִּרְכַּת הַגּוֹמֵל

הַחְזָרַת הַתּוֹרָה

סֵדֶר קְרִיאַת הַתּוֹרָה לְחוֹל

SEDER K'RIAT HATORAH L'CHOL — READING THE TORAH ON WEEKDAYS

The Ark is opened. Remove the Torah.
Standing with the Torah, recite these verses:

IN THIS SCROLL is the secret of our people's life from Sinai until now.
Its teaching is love and justice, goodness and hope.
Freedom is its gift to all who treasure it.

כִּי מִצִּיּוֹן תֵּצֵא תוֹרָה,
וּדְבַר־יְיָ מִירוּשָׁלָיִם.

FOR FROM OUT OF ZION will come the Torah,
and the word of Adonai from Jerusalem.

בָּרוּךְ שֶׁנָּתַן תּוֹרָה
לְעַמּוֹ יִשְׂרָאֵל בִּקְדֻשָּׁתוֹ.

BLESSED IS GOD who in holiness gave the Torah to the people Israel.

גַּדְּלוּ לַייָ אִתִּי,
וּנְרוֹמְמָה שְׁמוֹ יַחְדָּו.

EXALT ADONAI with me, let us extol God's name together.

לְךָ יְיָ הַגְּדֻלָּה וְהַגְּבוּרָה
וְהַתִּפְאֶרֶת וְהַנֵּצַח וְהַהוֹד,
כִּי כֹל בַּשָּׁמַיִם וּבָאָרֶץ.
לְךָ יְיָ הַמַּמְלָכָה
וְהַמִּתְנַשֵּׂא לְכֹל לְרֹאשׁ.

YOURS, Adonai, is the greatness, might, splendor, triumph, and majesty —
yes, all that is in heaven and on earth.
To You, Adonai belong sovereignty and preeminence above all.

The Torah is unwrapped.

כִּי מִצִּיּוֹן תֵּצֵא תוֹרָה *Ki miTziyon teitzei Torah . . . For from out of Zion . . .* Isaiah 2:3

גַּדְּלוּ לַייָ אִתִּי *Gadlu l'Adonai iti . . . Exalt Adonai with me . . .* Psalm 34:4

לְךָ יְיָ הַגְּדֻלָּה *L'cha Adonai hag'dulah . . . Yours, Adonai, is the greatness . . .* I Chronicles 29:11

Kabbalat HaTorah

Hakafah

Birchot HaTorah

Mi Shebeirach

Hagbahah

Birkat HaGomeil

Hachzarat HaTorah

HAKAFAH SELECTIONS

רוֹמְמוּ יְיָ אֱלֹהֵינוּ,
וְהִשְׁתַּחֲווּ לְהַר קָדְשׁוֹ,
כִּי קָדוֹשׁ יְיָ אֱלֹהֵינוּ.

EXALT ADONAI our God and bow down toward God's holy mountain,
for Adonai our God is holy.

עַל שְׁלֹשָׁה דְּבָרִים הָעוֹלָם עוֹמֵד:
עַל הַתּוֹרָה וְעַל הָעֲבוֹדָה
וְעַל גְּמִילוּת חֲסָדִים.

THE WORLD is sustained by three things: Torah, worship and loving deeds.

לֹא־יִשָּׂא גוֹי אֶל־גּוֹי חֶרֶב
וְלֹא־יִלְמְדוּ עוֹד מִלְחָמָה.

NATION SHALL not lift up sword against nation;
neither shall they learn war anymore.

הַלְלוּ . . .
כֹּל הַנְּשָׁמָה תְּהַלֵּל יָהּ,
הַלְלוּ, הַלְלוּ־יָהּ.

LET all that breathes praise God. Hallelujah!

רוֹמְמוּ יְיָ *Rom'mu Adonai . . . Exalt Adonai . . .* Psalm 99:9
עַל שְׁלֹשָׁה דְּבָרִים *Al sh'loshah d'varim . . . The world is sustained by three things . . . Pirkei Avot 1:2*
לֹא־יִשָּׂא גוֹי *Lo yisa goy . . . Nation shall not lift up . . .* Isaiah 2:4
כֹּל הַנְּשָׁמָה *Kol han'shamah . . . Let all that breathes . . .* Psalm 150:6

הָבוּ גֹדֶל לֵאלֹהֵינוּ,
וּתְנוּ כָבוֹד לַתּוֹרָה.

LET US DECLARE the greatness of our God and give honor to the Torah.

ONE WHO MAKES AN ALIYAH MIGHT OFFER:

MAY GOD be with you! — יְיָ עִמָּכֶם.

Congregation responds:

MAY GOD bless you! — יְבָרֶכְךָ יְיָ.

BLESSING BEFORE THE READING OF THE TORAH

בָּרְכוּ אֶת יְיָ הַמְבֹרָךְ.
בָּרוּךְ יְיָ הַמְבֹרָךְ לְעוֹלָם וָעֶד.
בָּרוּךְ אַתָּה, יְיָ
אֱלֹהֵינוּ, מֶלֶךְ הָעוֹלָם,
אֲשֶׁר בָּחַר בָּנוּ מִכָּל הָעַמִּים,
וְנָתַן לָנוּ אֶת תּוֹרָתוֹ.
בָּרוּךְ אַתָּה, יְיָ, נוֹתֵן הַתּוֹרָה.

BLESS ADONAI who is blessed.
Blessed is Adonai who is blessed now and forever.
Blessed are You, Adonai our God, Sovereign of the universe, who has chosen us from among the peoples, and given us the Torah. Blessed are You, Adonai, who gives the Torah.

BLESSING AFTER THE READING OF THE TORAH

בָּרוּךְ אַתָּה, יְיָ
אֱלֹהֵינוּ, מֶלֶךְ הָעוֹלָם,
אֲשֶׁר נָתַן לָנוּ תּוֹרַת אֱמֶת,
וְחַיֵּי עוֹלָם נָטַע בְּתוֹכֵנוּ.
בָּרוּךְ אַתָּה, יְיָ, נוֹתֵן הַתּוֹרָה.

BLESSED ARE YOU, Adonai our God, Sovereign of the universe,
who has given us a Torah of truth, implanting within us eternal life.
Blessed are You, Adonai, who gives the Torah.

T Proverbs 3:18 teaches, "It is a tree of life to all that hold fast to it, and all of its supporters are happy." This is understood to be a metaphor for the Torah. How does one hold fast to the Torah, and why are its supporters happy?

MI SHEBEIRACH FOR ALIYAH

מִי שֶׁבֵּרַךְ אֲבוֹתֵינוּ וְאִמּוֹתֵינוּ,
אַבְרָהָם יִצְחָק וְיַעֲקֹב,
שָׂרָה, רִבְקָה, רָחֵל וְלֵאָה,
הוּא יְבָרֵךְ אֶת *[name]* בֶּן/בַּת *[parents]*
בַּעֲבוּר שֶׁעָלָה/שֶׁעָלְתָה
לִכְבוֹד הַמָּקוֹם, וְלִכְבוֹד הַתּוֹרָה.
בִּשְׂכַר זֶה הַקָּדוֹשׁ בָּרוּךְ הוּא
יִשְׁמְרֵהוּ/יִשְׁמְרֶהָ
וְיַצִּילֵהוּ/וְיַצִּילֶהָ
מִכָּל צָרָה וְצוּקָה וּמִכָּל נֶגַע וּמַחֲלָה,
וְיִשְׁלַח בְּרָכָה וְהַצְלָחָה
בְּכָל מַעֲשֵׂה יָדָיו/יָדֶיהָ,
עִם כָּל יִשְׂרָאֵל. וְנֹאמַר: אָמֵן.

MAY THE ONE WHO BLESSED our ancestors Abraham, Isaac, and Jacob, Sarah, Rebecca, Rachel, and Leah, bless *[name]* son/daughter of *[parents]*, since he/she has come up to the Torah in honor of God and Torah. May he/she merit from the Holy One of Blessing protection, rescue from any trouble or distress, and from any illness, minor or serious; may God send blessing and success in his/her every endeavor, together with all Israel, and let us say, Amen.

HAGBAHAH UG'LILAH— הַגְבָּהָה וּגְלִילָה

The Torah is raised, rolled, and wrapped.

וְזֹאת הַתּוֹרָה אֲשֶׁר שָׂם מֹשֶׁה
לִפְנֵי בְּנֵי יִשְׂרָאֵל,
עַל־פִּי יְיָ בְּיַד־מֹשֶׁה.

THIS IS THE TORAH which Moses placed
before the people of Israel,
God's word through the hand of Moses.

Prayers of Our Community begin on page 112.

וְזֹאת הַתּוֹרָה *V'zot haTorah . . . This is the Torah . . .* Deuteronomy 4:44

עַל־פִּי יְיָ *al pi Adonai . . . God's word . . .* Numbers 9:23

PRAYERS FOR HEALING

מִי שֶׁבֵּרַךְ אֲבוֹתֵינוּ וְאִמּוֹתֵינוּ,
אַבְרָהָם, יִצְחָק וְיַעֲקֹב, שָׂרָה, רִבְקָה,
רָחֵל וְלֵאָה, הוּא יְבָרֵךְ אֶת הַחוֹלִים
[*names*]. הַקָּדוֹשׁ בָּרוּךְ הוּא יִמָּלֵא
רַחֲמִים עֲלֵיהֶם, לְהַחֲלִימָם וּלְרַפֹּאתָם
וּלְהַחֲזִיקָם, וְיִשְׁלַח לָהֶם מְהֵרָה
רְפוּאָה, רְפוּאָה שְׁלֵמָה מִן הַשָּׁמַיִם,
רְפוּאַת הַנֶּפֶשׁ וּרְפוּאַת הַגּוּף, הַשְׁתָּא
בַּעֲגָלָא וּבִזְמַן קָרִיב. וְנֹאמַר: אָמֵן.

MAY THE ONE who blessed our ancestors Abraham, Isaac, and Jacob, Sarah, Rebecca, Rachel, and Leah, bless and heal those who are ill [*names*]. May the Blessed Holy One be filled with compassion for their health to be restored and their strength to be revived. May God swiftly send them a complete renewal of body and spirit, and let us say, Amen.

מִי שֶׁבֵּרַךְ אֲבוֹתֵינוּ
מְקוֹר הַבְּרָכָה לְאִמּוֹתֵינוּ.

May the Source of strength who blessed the ones before us
help us find the courage to make our lives a blessing and let us say, Amen.

מִי שֶׁבֵּרַךְ אִמּוֹתֵינוּ
מְקוֹר הַבְּרָכָה לַאֲבוֹתֵינוּ.

Bless those in need of healing with *r'fuah sh'leimah,*
the renewal of body, the renewal of spirit, and let us say, Amen.

BIRKAT HAGOMEIL — בִּרְכַּת הַגּוֹמֵל — THANKSGIVING BLESSING

Individual recites:

בָּרוּךְ אַתָּה, יְיָ אֱלֹהֵינוּ,
מֶלֶךְ הָעוֹלָם, שֶׁגְּמָלַנִי כָּל טוֹב.

BLESSED ARE YOU, Adonai our God, Sovereign of the universe,
who has bestowed every goodness upon us.

Congregation responds:

אָמֵן. מִי שֶׁגְּמָלְכֶם כָּל טוֹב,
הוּא יִגְמָלְכֶם כָּל טוֹב. סֶלָה.

Amen. May the One who has bestowed goodness upon you
continue to bestow every goodness upon you forever.

בִּרְכַּת הַגּוֹמֵל *Birkat HaGomeil* — may be recited by one who has survived a life-challenging situation.

יְהַלְלוּ אֶת שֵׁם יְיָ,
כִּי נִשְׂגָּב שְׁמוֹ לְבַדּוֹ.

LET US PRAISE the Name of Adonai,
for God's Name alone is exalted!

הוֹדוֹ עַל אֶרֶץ וְשָׁמָיִם.
וַיָּרֶם קֶרֶן לְעַמּוֹ,
תְּהִלָּה לְכָל־חֲסִידָיו,
לִבְנֵי יִשְׂרָאֵל עַם־קְרֹבוֹ.
הַלְלוּ־יָהּ!

GOD'S MAJESTY is above the earth and heaven; and God is the strength of our people, making God's faithful ones, Israel, a people close to the Eternal. Halleluyah!

The Torah is returned to the Ark.

כִּי לֶקַח טוֹב נָתַתִּי לָכֶם,
תּוֹרָתִי אַל תַּעֲזֹבוּ.

עֵץ חַיִּים הִיא לַמַּחֲזִיקִים בָּהּ,
וְתֹמְכֶיהָ מְאֻשָּׁר.
דְּרָכֶיהָ דַרְכֵי נֹעַם,
וְכָל נְתִיבוֹתֶיהָ שָׁלוֹם.

הֲשִׁיבֵנוּ יְיָ אֵלֶיךָ וְנָשׁוּבָה,
חַדֵּשׁ יָמֵינוּ כְּקֶדֶם.

FOR I HAVE GIVEN YOU good instruction; do not abandon My Torah.

IT IS A TREE OF LIFE for those who hold fast to it, and all its supporters are happy.
Its ways are ways of pleasantness and all its paths are peace.
Return us to You, Adonai, and we will return; renew our days as of old.

הוֹדוֹ . . . יְהַלְלוּ *Y'hal'lu . . . Hodo . . . Let us praise . . . God's majesty* Psalm 148:13-14

כִּי לֶקַח טוֹב *Ki lekach tov . . . For I have given you . . . is an agglomeration of* Proverbs 4:2, Proverbs 3:18, Proverbs 3:17, *and* Lamentations 5:21

Kabbalat HaTorah
Hakafah
Birchot HaTorah
Mi Shebeirach
Hagbahah
Birkat HaGomeil
Hachzarat HaTorah

COMMUNITY

FOR OUR CONGREGATION

SOURCE of all being,
may the children of this community learn these passions from us:
love of Torah, devotion in prayer, and support of the needy.
May we guide with integrity, and may our leadership be in Your service.
May those who teach and nourish us be blessed with satisfaction,
and may we appreciate their time and their devotion.
Bless us with the fruits of wisdom and understanding,
and may our efforts bring fulfillment and joy.

בָּרוּךְ אַתָּה, יְיָ, שֶׁאוֹתְךָ לְבַדְּךָ בְּיִרְאָה נַעֲבוֹד.

FOR OUR COUNTRY

THUS SAYS ADONAI, This is what I desire:
to unlock the fetters of wickedness, and untie the cords of lawlessness;
to let the oppressed go free, to break off every yoke.
Share your bread with the hungry, and take the wretched poor into your home.
When you see the naked, give clothing, and do not ignore your own kin.

If you banish the yoke from your midst, the menacing hand, the evil speech;
if you offer compassion to the hungry and satisfy the famished creature —
then your light shall shine in darkness.

O GUARDIAN of life and liberty,
may our nation always merit Your protection.
Teach us to give thanks for what we have
by sharing it with those who are in need.
Keep our eyes open to the wonders of creation,
and alert to the care of the earth.
May we never be lazy in the work of peace;
may we honor those who have died in defense of our ideals.
Grant our leaders wisdom and forebearance.
May they govern with justice and compassion.
Help us all to appreciate one another,
and to respect the many ways that we may serve You.
May our homes be safe from affliction and strife,
and our country be sound in body and spirit.
Amen.

Thus says Adonai . . . Selected verses from Isaiah 58

PRAYERS OF OUR

FOR THE STATE OF ISRAEL

שַׁאֲלוּ שְׁלוֹם יְרוּשָׁלָיִם,
יִשְׁלָיוּ אֹהֲבָיִךְ.

PRAY for the peace of Jerusalem;
may those who love you prosper.

אָבִינוּ שֶׁבַּשָּׁמַיִם,
צוּר יִשְׂרָאֵל וְגוֹאֲלוֹ,
בָּרֵךְ אֶת מְדִינַת יִשְׂרָאֵל,
רֵאשִׁית צְמִיחַת גְּאֻלָּתֵנוּ.
הָגֵן עָלֶיהָ בְּאֶבְרַת חַסְדֶּךָ,
וּפְרוֹס עָלֶיהָ סֻכַּת שְׁלוֹמֶךָ.
וּשְׁלַח אוֹרְךָ וַאֲמִתְּךָ לְרָאשֶׁיהָ,
שָׂרֶיהָ וְיוֹעֲצֶיהָ,
וְתַקְּנֵם בְּעֵצָה טוֹבָה מִלְּפָנֶיךָ.
וְנָתַתָּ שָׁלוֹם בָּאָרֶץ,
וְשִׂמְחַת עוֹלָם לְיוֹשְׁבֶיהָ.
וְנֹאמַר: אָמֵן.

O HEAVENLY ONE, Protector and Redeemer of Israel,
bless the State of Israel which marks the dawning of hope for all who seek peace.
Shield it beneath the wings of Your love; spread over it the canopy of Your peace;
send Your light and truth to all who lead and advise,
guiding them with Your good counsel.
Establish peace in the land and fullness of joy for all who dwell there.
Amen.

שַׁאֲלוּ שְׁלוֹם יְרוּשָׁלָיִם *Shaalu sh'lom Y'rushalayim . . . Pray for the peace of Jerusalem . . .* Psalm 122:6

COMMUNITY

T'FILAT HADERECH — תְּפִלַּת הַדֶּרֶךְ — UPON SETTING FORTH ON A JOURNEY

יְהִי רָצוֹן מִלְּפָנֶיךָ יְיָ אֱלֹהֵינוּ
וֵאלֹהֵי אֲבוֹתֵינוּ וְאִמּוֹתֵינוּ,
שֶׁתּוֹלִיכֵנוּ לְשָׁלוֹם
וְתַעַזְרֵנוּ לְהַגִּיעַ לִמְחוֹז חֶפְצֵנוּ
לְחַיִּים וּלְשִׂמְחָה וּלְשָׁלוֹם.
וּשְׁמוֹר צֵאתֵנוּ וּבוֹאֵנוּ
וְתַצִּילֵנוּ מִכָּל צָרָה
וְתִשְׁלַח בְּרָכָה בְּכָל מַעֲשֵׂי יָדֵינוּ,
וּמַעֲשֵׂינוּ יְכַבְּדוּ אֶת שְׁמֶךָ.
בָּרוּךְ אַתָּה, יְיָ, שׁוֹמֵר יִשְׂרָאֵל לָעַד.

MAY IT BE YOUR WILL, our God and God of our ancestors,
that You lead us in peace and help us reach our destination
safely, joyfully and peacefully.
May You protect us on our leaving and on our return,
and rescue us from any harm,
and may You bless the work of our hands,
and may our deeds merit honor for You.
Praise to You, Adonai, Protector of Israel.

בָּרוּךְ אַתָּה, יְיָ, שׁוֹמֵר יִשְׂרָאֵל לָעַד.

PRAYERS OF OUR

L'ROSH CHODESH — לְרֹאשׁ חֹדֶשׁ — FOR THE NEW MONTH

יְהִי רָצוֹן מִלְּפָנֶיךָ,
יְיָ אֱלֹהֵינוּ
וֵאלֹהֵי אֲבוֹתֵינוּ וְאִמּוֹתֵינוּ,
שֶׁתְּחַדֵּשׁ עָלֵינוּ אֶת הַחֹדֶשׁ
הַבָּא (הַזֶּה) לְטוֹבָה וְלִבְרָכָה.
וְתִתֶּן לָנוּ חַיִּים אֲרֻכִּים,
חַיִּים שֶׁל שָׁלוֹם,
חַיִּים שֶׁל פַּרְנָסָה,
חַיִּים שֶׁתְּהֵא בָנוּ
אַהֲבַת תּוֹרָה וְיִרְאַת שָׁמַיִם,
חַיִּים שֶׁיִּמָּלְאוּ מִשְׁאֲלוֹת
לִבֵּנוּ לְטוֹבָה. אָמֵן.

OUR GOD and God of our ancestors,
may the new month bring us goodness and blessing.
May we have long life, peace, prosperity,
a life exalted by love of Torah and reverence for the divine;
a life in which the longings of our hearts are fulfilled for good.

רֹאשׁ חֹדֶשׁ *[name of month]* יִהְיֶה בְּיוֹם *[day]* / הוּא הַיּוֹם

THE NEW MONTH of ______ will begin on ______ / begins today.

The custom of announcing the beginning of the new Hebrew month in the synagogue dates to the geonic period (circa 9th century). This prayer is recited on the Shabbat preceding the beginning of the new Hebrew month, with the exception of the month of Tishrei, which always coincides with Rosh HaShanah. This Shabbat is called Shabbat M'vorchim. We pray for blessing in the month ahead.

COMMUNITY

FOR A BAR AND BAT MITZVAH

INTO OUR HANDS, O God, You have placed Your Torah, to be held high by parents and children, and taught by one generation to the next. Whatever has befallen us, our people have remained steadfast in loyalty to the Torah. It was carried in the arms of parents that their children might not be deprived of their birthright.

And now, we pray that you, [*name*], may always be worthy of this inheritance. Take its teaching into your heart, and in turn pass it on to your children and those who come after you. May you be a faithful Jew, searching for wisdom and truth, working for justice and peace.

May the One who has always been our Guide inspire you to bring honor to our family and to the House of Israel.

בָּרוּךְ אַתָּה, יְיָ, שֶׁנָּתַן לִי אֶת הַזְּכוּת וְאֶת הַכָּבוֹד לָתֵת לְךָ/לָךְ תּוֹרָה.

Blessed is Adonai our God, who gives me the honor and privilege
of entrusting you with Torah.

OUR HEARTS are one on this joyous day
as you commit yourself to a life of Torah:
a life, we pray, filled with
wisdom, caring and right action.

We pray that you will grow each day
in compassion for the needy,
in concern for the stranger,
in love of all people.

May the One who blessed our ancestors,
Abraham and Sarah, Isaac and Rebecca,
Jacob and Rachel and Leah,
bless you on your becoming a Bar/t Mitzvah.

May you grow with strength and courage,
with vision and sensitivity.
And may you always be certain of our love.

Amen.

PRAYERS OF OUR

E Individual communities write their own prayers. What is your prayer or blessing for the community with which you are worshipping?

תְּפִלּוֹת לְשַׁבָּת

T'FILOT L'SHABBAT

PRAYERS FOR SHABBAT

KABBALAT PANIM — WELCOMING

BLESSINGS FOR SHABBAT

The candles are lit before the blessing is recited.

בָּרוּךְ אַתָּה, יְיָ
אֱלֹהֵינוּ, מֶלֶךְ הָעוֹלָם,
אֲשֶׁר קִדְּשָׁנוּ בְּמִצְוֹתָיו,
וְצִוָּנוּ לְהַדְלִיק
נֵר שֶׁל שַׁבָּת.

BLESSED ARE YOU, Adonai our God, Sovereign of the universe,
who hallows us with mitzvot,
commanding us to kindle the light of Shabbat.

For Kabbalat Shabbat, turn to page 130.

Shabbat Candle Blessing — The mitzvah of kindling Shabbat lights in the home is an early rabbinic practice (*M. Shabbat 2:1ff.*) The Shabbat candle blessing is first recorded in the ninth-century prayerbook, *Seder Rav Amram*. Lighting Shabbat candles as part of the synagogue service is an innovation of Reform Judaism.

I Close your eyes and imagine a world without light. Far in the distance, see the Shabbat candles. As you move toward them, think about how the presence of the light changes the way you view the darkness. If Shabbat is a light at the end of the week, how does it illumine everyday life?

E When do you mark a significant moment — a time when you move from one experience to another — with a ritual? How does that ritual fix that moment in time and memory?

KIDDUSH FOR EVENING OF SHABBAT

Fill a Kiddush cup with wine or grape juice.
Raise it and recite:

וַיְהִי עֶרֶב וַיְהִי בֹקֶר
יוֹם הַשִּׁשִּׁי.

AND THERE WAS EVENING and there was morning,
the sixth day.

וַיְכֻלּוּ הַשָּׁמַיִם וְהָאָרֶץ
וְכָל־צְבָאָם.
וַיְכַל אֱלֹהִים בַּיּוֹם הַשְּׁבִיעִי
מְלַאכְתּוֹ אֲשֶׁר עָשָׂה.
וַיִּשְׁבֹּת בַּיּוֹם הַשְּׁבִיעִי
מִכָּל־מְלַאכְתּוֹ אֲשֶׁר עָשָׂה.
וַיְבָרֶךְ אֱלֹהִים אֶת־יוֹם הַשְּׁבִיעִי
וַיְקַדֵּשׁ אֹתוֹ, כִּי בוֹ שָׁבַת מִכָּל־
מְלַאכְתּוֹ אֲשֶׁר בָּרָא אֱלֹהִים לַעֲשׂוֹת.

THE HEAVEN AND THE EARTH were finished, and all their array.
On the seventh day God finished the work that God had been doing,
and God ceased on the seventh day from all the work that God had done.
And God blessed the seventh day and declared it holy,
because on it God ceased from all the work of creation that God had done.

Shabbat Kiddush — The beginning of Shabbat is marked by reciting a benediction sanctifying the day (*Kiddush HaYom* or *K'dushat HaYom,* "Sanctification of the Day"; *M. B'rachot* 8:1 and *P'sachim* 10:2). The benediction praises God for the gift of Shabbat that marks Israel as unique. Since Kiddush is recited over a cup of wine, symbolizing joy, it is preceded by the wine benediction, *borei p'ri hagafen,* "Creator of the fruit of the vine."

וַיְהִי עֶרֶב *Vay'hi erev . . . And there was evening . . .* Genesis 1:31

וַיְכֻלּוּ *Vay'chulu . . . The heaven and the earth . . .* Genesis 2:1–3

בָּרוּךְ אַתָּה, יְיָ
אֱלֹהֵינוּ, מֶלֶךְ הָעוֹלָם,
בּוֹרֵא פְּרִי הַגָּפֶן.

בָּרוּךְ אַתָּה, יְיָ
אֱלֹהֵינוּ, מֶלֶךְ הָעוֹלָם,
אֲשֶׁר קִדְּשָׁנוּ בְּמִצְוֹתָיו וְרָצָה בָנוּ,
וְשַׁבַּת קָדְשׁוֹ
בְּאַהֲבָה וּבְרָצוֹן הִנְחִילָנוּ,
זִכָּרוֹן לְמַעֲשֵׂה בְרֵאשִׁית.
כִּי הוּא יוֹם תְּחִלָּה לְמִקְרָאֵי קֹדֶשׁ,
זֵכֶר לִיצִיאַת מִצְרָיִם.
כִּי בָנוּ בָחַרְתָּ, וְאוֹתָנוּ קִדַּשְׁתָּ
מִכָּל הָעַמִּים.
וְשַׁבַּת קָדְשְׁךָ
בְּאַהֲבָה וּבְרָצוֹן הִנְחַלְתָּנוּ.
בָּרוּךְ אַתָּה, יְיָ, מְקַדֵּשׁ הַשַּׁבָּת.

PRAISE TO YOU, Adonai our God, Sovereign of the universe,
Creator of the fruit of the vine.

Praise to You, Adonai our God, Sovereign of the universe
who finding favor with us, sanctified us with mitzvot.
In love and favor, You made the holy Shabbat our heritage
as a reminder of the work of Creation.
As first among our sacred days, it recalls the Exodus from Egypt.
You chose us and set us apart from the peoples.
In love and favor You have given us Your holy Shabbat as an inheritance.

Praise to You, Adonai, who sanctifies Shabbat.

בָּרוּךְ אַתָּה, יְיָ, מְקַדֵּשׁ הַשַּׁבָּת.

WE ENTER THIS SANCTUARY to welcome Shabbat.
Within these walls we sit surrounded by numberless generations.
Our ancestors built the synagogue as a visible sign of God's Presence in their midst.
Throughout our long history and our endless wanderings, it has endured,
a beacon of truth, love, and justice for all humanity.
Its presence guided our ancestors to lives of righteousness,
holding up to them a vision of their truest selves.

Now we, in our turn, come into this sanctuary to affirm the sacredness of our lives.
May we enter this place in peace.
May holiness wrap around us as we cross its threshold.
Weariness, doubt, the flaws within our human hearts,
the harshness of the week — let these drop away at the door.
In the brightness of Shabbat, let peace settle upon us as we lift our hearts in prayer.

MAY THE DOOR of this synagogue be wide enough
to receive all who hunger for love, all who are lonely for friendship.

May it welcome all who have cares to unburden,
thanks to express, hopes to nurture.

May the door of this synagogue be narrow enough
to shut out pettiness and pride, envy and enmity.

May its threshold be no stumbling block
to young or straying feet.

May it be too high to admit complacency,
selfishness and harshness.

May this synagogue be, for all who enter,
the doorway to a richer and more meaningful life.

Candle Blessing
Kiddush, Evening
Welcome
Shabbat Songs

I BEGIN WITH A PRAYER of gratitude
for all that is holy in my life.
God needs no words, no English or Hebrew,
no semantics and no services.
But I need them.
Through prayer, I can sense my inner strength,
my inner purpose,
my inner joy, my capacity to love.
As I reach upward in prayer,
I sense these qualities in my Creator.
To love God is to love each other,
to work to make our lives better.
To love God is to love the world God created
and to work to perfect it.
To love God is to love dreams of peace and joy
that illumine all of us,
and to bring that vision to life.

WE OFFER THANKS, O God, for this Shabbat
which unites us in faith and hope.

For Shabbat holiness, which inspires sacred living,
for Shabbat memories, glowing even in darkness,
for Shabbat peace, born of friendship and love,
we offer thanks and blessing, O God.

Candle Blessing
Kiddush, Evening
Welcome
Shabbat Songs

HINEIH MAH TOV

הִנֵּה מַה־טּוֹב וּמַה־נָּעִים
שֶׁבֶת אַחִים גַּם־יָחַד.

How good and how pleasant it is that brothers and sisters dwell together.
(Psalm 133:1)

MAH YAFEH HAYOM

מַה יָּפֶה הַיּוֹם, שַׁבָּת שָׁלוֹם.

How lovely today is, Shabbat Shalom.

Y'DID NEFESH

יְדִיד נֶפֶשׁ, אָב הָרַחֲמָן,
מְשׁוֹךְ עַבְדְּךָ אֶל רְצוֹנֶךָ.
יָרוּץ עַבְדְּךָ כְּמוֹ אַיָּל,
יִשְׁתַּחֲוֶה אֶל מוּל הֲדָרֶךָ.

Heart's delight, Source of mercy, draw Your servant into Your arms:
I leap like a deer to stand in awe before You.

SHABBAT HAMALKAH

הַחַמָּה מֵרֹאשׁ הָאִילָנוֹת נִסְתַּלְּקָה,
בֹּאוּ וְנֵצֵא לִקְרַאת שַׁבָּת הַמַּלְכָּה.
הִנֵּה הִיא יוֹרֶדֶת, הַקְּדוֹשָׁה הַבְּרוּכָה.
וְעִמָּהּ מַלְאָכִים, צְבָא שָׁלוֹם וּמְנוּחָה.
בֹּאִי בֹּאִי הַמַּלְכָּה, בֹּאִי בֹּאִי הַכַּלָּה.
שָׁלוֹם עֲלֵיכֶם, מַלְאֲכֵי הַשָּׁלוֹם.

The sun on the treetops no longer is seen;
come, gather to welcome the Sabbath, our queen.
Behold her descending, the holy, the blessed,
and with her the angels of peace and of rest.
Draw near, draw near, and here abide,
draw near, draw near, O Sabbath bride.
Peace also to you, you angels of peace.

DODI LI

דּוֹדִי לִי וַאֲנִי לוֹ הָרֹעֶה בַּשּׁוֹשַׁנִּים.
מִי זֹאת עֹלָה מִן־הַמִּדְבָּר,
מְקֻטֶּרֶת מוֹר וּלְבוֹנָה . . .
לִבַּבְתִּנִי, אֲחֹתִי כַלָּה.
עוּרִי צָפוֹן וּבוֹאִי תֵימָן . . .

My beloved is mine and I am my beloved's who browses among the lilies.
Who is this that comes up from the desert, in clouds of myrrh and frankincense?
You have captured my heart, my sister, my bride.
Awake, O north wind, come, O south wind!
(Song of Songs 2:16; 3:6; 4:9, 16)

KOL DODI

קוֹל דּוֹדִי הִנֵּה זֶה־בָּא,
מְדַלֵּג עַל־הֶהָרִים,
מְקַפֵּץ עַל־הַגְּבָעוֹת.

Hark! My beloved comes leaping over the mountains, bounding over the hills.
(Song of Songs 2:8)

KI ESHM'RAH SHABBAT

כִּי אֶשְׁמְרָה שַׁבָּת אֵל יִשְׁמְרֵנִי.
אוֹת הִיא לְעוֹלְמֵי עַד בֵּינוֹ וּבֵינִי.

When I keep Shabbat, God watches over me. It is a sign forever between God and me.

EILEH CHAMDAH LIBI

אֵלֶּה חָמְדָה לִבִּי,
חוּסָה נָּא וְאַל נָא תִּתְעַלָּם.

God is my heart's desire. Appear! Do not hide.

קַבָּלַת שַׁבָּת

KABBALAT SHABBAT — WELCOMING SHABBAT

PSALM 95:1-7

לְכוּ נְרַנְּנָה לַייָ,
נָרִֽיעָה לְצוּר יִשְׁעֵֽנוּ.
נְקַדְּמָה פָנָיו בְּתוֹדָה,
בִּזְמִרוֹת נָרִֽיעַ לוֹ.

כִּי אֵל גָּדוֹל יְיָ,
וּמֶֽלֶךְ גָּדוֹל עַל־כָּל־אֱלֹהִים.
אֲשֶׁר בְּיָדוֹ מֶחְקְרֵי־אָֽרֶץ,
וְתוֹעֲפוֹת הָרִים לוֹ.
אֲשֶׁר־לוֹ הַיָּם וְהוּא עָשָֽׂהוּ,
וְיַבֶּֽשֶׁת יָדָיו יָצָֽרוּ.

בֹּֽאוּ נִשְׁתַּחֲוֶה וְנִכְרָֽעָה
נִבְרְכָה לִפְנֵי־יְיָ עֹשֵֽׂנוּ.
כִּי־הוּא אֱלֹהֵֽינוּ,
וַאֲנַֽחְנוּ עַם מַרְעִיתוֹ וְצֹאן יָדוֹ,

הַיּוֹם אִם־בְּקֹלוֹ תִשְׁמָֽעוּ.

COME, LET US SING joyously to Adonai,
raise a shout for our Rock and Deliverer;
let us come into God's presence with praise;
let us raise a shout for God in song!
For Adonai is a great God,
the great ruler of all divine beings.

קַבָּלַת שַׁבָּת *Kabbalat Shabbat,* "Welcoming Shabbat," is one of the ritual innovations of the circle of Kabbalists (Jewish mystics) who gathered around Rabbi Isaac Luria in sixteenth-century Safed in the Land of Israel.

Psalms 95–99, 29 — This cycle of six psalms, representing the six days of creation, enjoins all creation to sing out the praises of God. God's creative power is revealed in the world around us; God's sovereignty is over all nature and all nations.

In God's hand are the depths of the earth;
the peaks of the mountains are God's.
God's is the sea, God made it;
and the land, which God's hands fashioned.
Come, let us bow down and kneel,
bend the knee before Adonai our maker,
for Adonai is our God,
and we are the people God tends, the flock in God's care.
O, if you would but heed God's charge this day.

PSALM 96:1-6, 11-13

שִׁירוּ לַייָ שִׁיר חָדָשׁ,
שִׁירוּ לַייָ כָּל־הָאָֽרֶץ.
שִׁירוּ לַייָ, בָּרְכוּ שְׁמוֹ,
בַּשְּׂרוּ מִיּוֹם־לְיוֹם יְשׁוּעָתוֹ.
סַפְּרוּ בַגּוֹיִם כְּבוֹדוֹ,
בְּכָל־הָעַמִּים נִפְלְאוֹתָיו.
כִּי גָדוֹל יְיָ וּמְהֻלָּל מְאֹד,
נוֹרָא הוּא עַל־כָּל־אֱלֹהִים.
כִּי כָּל־אֱלֹהֵי הָעַמִּים אֱלִילִים
וַייָ שָׁמַֽיִם עָשָׂה.
הוֹד־וְהָדָר לְפָנָיו,
עֹז וְתִפְאֶֽרֶת בְּמִקְדָּשׁוֹ.

יִשְׂמְחוּ הַשָּׁמַֽיִם וְתָגֵל הָאָֽרֶץ
יִרְעַם הַיָּם וּמְלֹאוֹ.
יַעֲלֹז שָׂדַי וְכָל־אֲשֶׁר־בּוֹ,
אָז יְרַנְּנוּ כָּל־עֲצֵי־יָֽעַר.
לִפְנֵי יְיָ כִּי בָא, כִּי בָא לִשְׁפֹּט הָאָֽרֶץ,
יִשְׁפֹּט תֵּבֵל בְּצֶֽדֶק וְעַמִּים בֶּאֱמוּנָתוֹ.

מִזְמוֹר צ"ה-צ"ט, כ"ט

לְכָה דוֹדִי

מִזְמוֹר צ"ב-צ"ג

שָׁלוֹם עֲלֵיכֶם

Sɪɴɢ ᴛᴏ Aᴅᴏɴᴀɪ a new song,
sing to Adonai, all the earth.
Sing to Adonai, bless God's name,
proclaim God's victory day after day.
Tell of God's glory among the nations,
God's wondrous deeds, among all peoples.
For Adonai is great and much acclaimed,
God is held in awe by all divine beings.
All the gods of the peoples are mere idols,
but Adonai made the heavens.
Glory and majesty are before God;
strength and splendor are in God's temple.
   Let the heavens rejoice and the earth exult;
   let the sea and all within it thunder,
   the fields and everything in them exult;
   then shall all the trees of the forest shout for joy
   at the presence of Adonai, for God is coming,
   for God is coming to rule the earth;
   God will rule the world justly,
   and its peoples in faithfulness.

PSALM 97:1-2, 10-12

יְיָ מָלָךְ, תָּגֵל הָאָרֶץ,
יִשְׂמְחוּ אִיִּים רַבִּים.
עָנָן וַעֲרָפֶל סְבִיבָיו
צֶדֶק וּמִשְׁפָּט מְכוֹן כִּסְאוֹ.

אֹהֲבֵי יְיָ שִׂנְאוּ רָע,
שֹׁמֵר נַפְשׁוֹת חֲסִידָיו
מִיַּד רְשָׁעִים יַצִּילֵם.

אוֹר זָרֻעַ לַצַּדִּיק
וּלְיִשְׁרֵי־לֵב שִׂמְחָה.
שִׂמְחוּ צַדִּיקִים בַּיְיָ
וְהוֹדוּ לְזֵכֶר קָדְשׁוֹ.

ADONAI IS SOVEREIGN!
Let the earth exult, the many islands rejoice!
Dense clouds are around God,
righteousness and justice are the base of God's throne.
O you who love Adonai, hate evil!
God guards the lives of God's loyal ones,
saving them from the hand of the wicked.
Light is sown for the righteous,
radiance for the upright.
O you righteous, rejoice in Adonai
and acclaim God's holy name!

PSALM 98:1-9

מִזְמוֹר.
שִׁירוּ לַייָ שִׁיר חָדָשׁ,
כִּי־נִפְלָאוֹת עָשָׂה,
הוֹשִׁיעָה־לּוֹ יְמִינוֹ וּזְרוֹעַ קָדְשׁוֹ.
הוֹדִיעַ יְיָ יְשׁוּעָתוֹ,
לְעֵינֵי הַגּוֹיִם גִּלָּה צִדְקָתוֹ.
זָכַר חַסְדּוֹ וֶאֱמוּנָתוֹ
לְבֵית יִשְׂרָאֵל, רָאוּ כָל־אַפְסֵי־אָרֶץ
אֵת יְשׁוּעַת אֱלֹהֵינוּ.

הָרִיעוּ לַייָ כָּל־הָאָרֶץ
פִּצְחוּ וְרַנְּנוּ וְזַמֵּרוּ.

A PSALM.
Sing to Adonai a new song,
for God has worked wonders;
God's right hand, God's holy arm,
has won God victory.
Adonai has manifested God's victory,
has displayed God's triumph in the sight of the nations.
God was mindful of God's steadfast love and faithfulness
toward the house of Israel;
all the ends of the earth beheld the victory of our God.
Raise a shout to Adonai, all the earth,
break into joyous songs of praise!

מִזְמוֹר צ"ה-צ"ט, כ"ט
לְכָה דוֹדִי
מִזְמוֹר צ"ב-צ"ג
שָׁלוֹם עֲלֵיכֶם

זַמְּרוּ לַייָ בְּכִנּוֹר,
בְּכִנּוֹר וְקוֹל זִמְרָה.
בַּחֲצֹצְרוֹת וְקוֹל שׁוֹפָר
הָרִיעוּ לִפְנֵי הַמֶּלֶךְ יְיָ.
יִרְעַם הַיָּם וּמְלֹאוֹ
תֵּבֵל וְיֹשְׁבֵי בָהּ.
נְהָרוֹת יִמְחֲאוּ־כָף
יַחַד הָרִים יְרַנֵּנוּ.
לִפְנֵי־יְיָ כִּי־בָא
לִשְׁפֹּט הָאָרֶץ
יִשְׁפֹּט־תֵּבֵל בְּצֶדֶק
וְעַמִּים בְּמֵישָׁרִים.

SING PRAISE TO ADONAI with the lyre,
with the lyre and melodious song.
With trumpets and the blast of the horn
raise a shout before Adonai, the ruler.
Let the sea and all within it thunder,
the world and its inhabitants;
let the rivers clap their hands,
the mountains sing joyously together
at the presence of Adonai,
for God is coming to rule the earth;
God will rule the world justly,
and its peoples with equity.

PSALM 99: 1-5, 9

יְיָ מָלָךְ יִרְגְּזוּ עַמִּים,
יֹשֵׁב כְּרוּבִים תָּנוּט הָאָרֶץ.
יְיָ בְּצִיּוֹן גָּדוֹל
וְרָם הוּא עַל־כָּל־הָעַמִּים.
יוֹדוּ שִׁמְךָ גָּדוֹל וְנוֹרָא,
קָדוֹשׁ הוּא.

וְעֹז מֶלֶךְ מִשְׁפָּט אָהֵב,
אַתָּה כּוֹנַנְתָּ מֵישָׁרִים,
מִשְׁפָּט וּצְדָקָה בְּיַעֲקֹב
אַתָּה עָשִׂיתָ.
רוֹמְמוּ יְיָ אֱלֹהֵינוּ
וְהִשְׁתַּחֲווּ לַהֲדֹם רַגְלָיו,
קָדוֹשׁ הוּא.

רוֹמְמוּ יְיָ אֱלֹהֵינוּ
וְהִשְׁתַּחֲווּ לְהַר קָדְשׁוֹ,
כִּי־קָדוֹשׁ יְיָ אֱלֹהֵינוּ.

ADONAI ENTHRONED on cherubim, is sovereign,
peoples tremble, the earth quakes.
Adonai is great in Zion,
and exalted above all peoples.
They praise Your name as great and awesome;
God is holy!
Mighty ruler who loves justice,
it was You who established equity,
You who worked righteous judgment in Jacob.
Exalt Adonai our God
and bow down to God's footstool;
God is holy!
  Exalt Adonai our God,
  and bow toward God's holy hill,
  for Adonai our God is holy.

מִזְמוֹר צ"ה-צ"ט, כ"ט

לְכָה דוֹדִי

מִזְמוֹר צ"ב-צ"ג

שָׁלוֹם עֲלֵיכֶם

Psalms 95-99, 29

L'chah Dodi

Psalms 92-93

Shalom Aleichem

PSALM 29:1-11

מִזְמוֹר לְדָוִד.
הָבוּ לַייָ בְּנֵי אֵלִים,
הָבוּ לַייָ כָּבוֹד וָעֹז.
הָבוּ לַייָ כְּבוֹד שְׁמוֹ,
הִשְׁתַּחֲווּ לַייָ בְּהַדְרַת־קֹדֶשׁ.
קוֹל יְיָ עַל־הַמָּיִם,
אֵל־הַכָּבוֹד הִרְעִים,
יְיָ עַל־מַיִם רַבִּים.
קוֹל־יְיָ בַּכֹּחַ קוֹל יְיָ בֶּהָדָר.
קוֹל יְיָ שֹׁבֵר אֲרָזִים
וַיְשַׁבֵּר יְיָ אֶת־אַרְזֵי הַלְּבָנוֹן,
וַיַּרְקִידֵם כְּמוֹ־עֵגֶל
לְבָנוֹן וְשִׂרְיוֹן כְּמוֹ בֶן־רְאֵמִים.

A PSALM OF DAVID.
Ascribe to Adonai, O divine beings,
ascribe to Adonai glory and strength.
Ascribe to Adonai the glory of God's name;
bow down to Adonai, majestic in holiness.
The voice of Adonai is over the waters;
the God of glory thunders,
Adonai, over the mighty waters.
The voice of Adonai is power;
the voice of Adonai is majesty;
the voice of Adonai breaks cedars;
Adonai shatters the cedars of Lebanon.
God makes Lebanon skip like a calf,
Sirion, like a young wild ox.

קוֹל־יְיָ חֹצֵב לַהֲבוֹת אֵשׁ,
קוֹל יְיָ יָחִיל מִדְבָּר,
יָחִיל יְיָ מִדְבַּר קָדֵשׁ.
קוֹל יְיָ יְחוֹלֵל אַיָּלוֹת
וַיֶּחֱשֹׂף יְעָרוֹת
וּבְהֵיכָלוֹ כֻּלּוֹ אֹמֵר כָּבוֹד.
יְיָ לַמַּבּוּל יָשָׁב
וַיֵּשֶׁב יְיָ מֶלֶךְ לְעוֹלָם.

יְיָ עֹז לְעַמּוֹ יִתֵּן,
יְיָ יְבָרֵךְ אֶת־עַמּוֹ בַשָּׁלוֹם.

THE VOICE OF ADONAI kindles flames of fire;
the voice of Adonai convulses the wilderness;
Adonai convulses the wilderness of Kadesh;
the voice of Adonai causes hinds to calve,
and strips forests bare;
while in God's temple all say "Glory!"
Adonai sat enthroned at the Flood;
Adonai sits enthroned, sovereign forever.
May Adonai grant strength to God's people;
may Adonai bestow on God's people wellbeing.

מִזְמוֹר צ"ה-צ"ט, כ"ט

לְכָה דוֹדִי

מִזְמוֹר צ"ב-צ"ג

שָׁלוֹם עֲלֵיכֶם

לְכָה דוֹדִי לִקְרַאת כַּלָּה,
פְּנֵי שַׁבָּת נְקַבְּלָה.

BELOVED, COME to meet the bride; beloved come to greet Shabbat.

1. שָׁמוֹר וְזָכוֹר בְּדִבּוּר אֶחָד,
הִשְׁמִיעָנוּ אֵל הַמְיֻחָד,
יְיָ אֶחָד וּשְׁמוֹ אֶחָד,
לְשֵׁם וּלְתִפְאֶרֶת וְלִתְהִלָּה.

"Keep" and "remember": a single command the Only God caused us to hear;
the Eternal is One, God's Name is One; glory and praise are God's.

2. לִקְרַאת שַׁבָּת לְכוּ וְנֵלְכָה,
כִּי הִיא מְקוֹר הַבְּרָכָה,
מֵרֹאשׁ מִקֶּדֶם נְסוּכָה,
סוֹף מַעֲשֶׂה בְּמַחֲשָׁבָה תְּחִלָּה.

Come with me to meet Shabbat, forever a fountain of blessing.
Still it flows, as from the start: the last of days, for which the first was made.

3. מִקְדַּשׁ מֶלֶךְ עִיר מְלוּכָה,
קוּמִי צְאִי מִתּוֹךְ הַהֲפֵכָה,
רַב לָךְ שֶׁבֶת בְּעֵמֶק הַבָּכָא,
וְהוּא יַחֲמוֹל עָלַיִךְ חֶמְלָה.

Royal shrine, city of kings, rise up and leave your ravaged state.
You have dwelt long enough in the valley of tears; now God will shower mercy on you.

4. הִתְנַעֲרִי, מֵעָפָר קוּמִי,
לִבְשִׁי בִּגְדֵי תִפְאַרְתֵּךְ, עַמִּי,
עַל יַד בֶּן יִשַׁי בֵּית הַלַּחְמִי,
קָרְבָה אֶל נַפְשִׁי גְאָלָהּ.

Lift yourself up! Shake off the dust! Array yourself in beauty, O my people!
At hand is Bethlehem's David, Jesse's son, bringing deliverance into my life.

This poem was composed by Shlomo Halevi Alkabetz, 16th century Safed kabbalist. The first eight verses are arranged acrostically according to the author's name (שְׁלֹמֹה הַלֵּוִי).

Psalms 95-99, 29
L'chah Dodi
Psalms 92-93
Shalom Aleichem

5. הִתְעוֹרְרִי, הִתְעוֹרְרִי,
כִּי בָא אוֹרֵךְ, קוּמִי אוֹרִי,
עוּרִי עוּרִי, שִׁיר דַּבֵּרִי,
כְּבוֹד יְיָ עָלַיִךְ נִגְלָה.

Awake, awake, your light has come! Arise, shine, awake and sing:
the Eternal's glory dawns upon you.

6. לֹא תֵבֹשִׁי וְלֹא תִכָּלְמִי,
מַה תִּשְׁתּוֹחֲחִי וּמַה תֶּהֱמִי,
בָּךְ יֶחֱסוּ עֲנִיֵּי עַמִּי,
וְנִבְנְתָה עִיר עַל תִּלָּהּ.

An end to shame and degradation; forget your sorrow; quiet your groans. The afflicted of my people find respite in you, the city renewed upon its ancient ruins.

7. וְהָיוּ לִמְשִׁסָּה שֹׁאסָיִךְ,
וְרָחֲקוּ כָּל מְבַלְּעָיִךְ,
יָשִׂישׂ עָלַיִךְ אֱלֹהָיִךְ,
כִּמְשׂוֹשׂ חָתָן עַל כַּלָּה.

The scavengers are scattered, your devourers have fled;
as a bridegroom rejoices in his bride, your God takes joy in you.

8. יָמִין וּשְׂמֹאל תִּפְרוֹצִי,
וְאֶת־יְיָ תַּעֲרִיצִי,
עַל יַד אִישׁ בֶּן פַּרְצִי,
וְנִשְׂמְחָה וְנָגִילָה.

Your space will be broad, your worship free: await the promised one;
we will exult, we will sing for joy!

9. בּוֹאִי בְשָׁלוֹם עֲטֶרֶת בַּעְלָהּ,
גַּם בְּשִׂמְחָה וּבְצָהֳלָה,
תּוֹךְ אֱמוּנֵי עַם סְגֻלָּה,
בּוֹאִי כַלָּה, בּוֹאִי כַלָּה.

Enter in peace, O crown of your husband; enter in gladness, enter in joy.
Come to the people that keeps its faith. Enter, O bride! Enter, O bride!

For those who choose: When the congregation reaches the last verse, beginning בּוֹאִי בְשָׁלוֹם *Bo-i v'shalom, Enter in peace,* all rise and turn toward the entrance of the sanctuary, as if to greet the Presence of Shabbat.

מִזְמוֹר צ"ה-צ"ט, כ"ט

לְכָה דוֹדִי

מִזְמוֹר צ"ב-צ"ג

שָׁלוֹם עֲלֵיכֶם

PSALM 92:1–7, 13–16

מִזְמוֹר שִׁיר
לְיוֹם הַשַּׁבָּת.
טוֹב לְהֹדוֹת לַיָי
וּלְזַמֵּר לְשִׁמְךָ עֶלְיוֹן.
לְהַגִּיד בַּבֹּקֶר חַסְדֶּךָ,
וֶאֱמוּנָתְךָ בַּלֵּילוֹת.
עֲלֵי־עָשׂוֹר וַעֲלֵי־נָבֶל,
עֲלֵי הִגָּיוֹן בְּכִנּוֹר.
כִּי שִׂמַּחְתַּנִי יָי בְּפָעֳלֶךָ
בְּמַעֲשֵׂי יָדֶיךָ אֲרַנֵּן.
מַה־גָּדְלוּ מַעֲשֶׂיךָ, יָי,
מְאֹד עָמְקוּ מַחְשְׁבֹתֶיךָ.
אִישׁ־בַּעַר לֹא יֵדָע
וּכְסִיל לֹא־יָבִין אֶת־זֹאת.
צַדִּיק כַּתָּמָר יִפְרָח
כְּאֶרֶז בַּלְּבָנוֹן יִשְׂגֶּה.
שְׁתוּלִים בְּבֵית יָי,
בְּחַצְרוֹת אֱלֹהֵינוּ יַפְרִיחוּ.
עוֹד יְנוּבוּן בְּשֵׂיבָה,
דְּשֵׁנִים וְרַעֲנַנִּים יִהְיוּ.
לְהַגִּיד כִּי־יָשָׁר יָי,
צוּרִי וְלֹא־עַוְלָתָה בּוֹ.

A PSALM. A SONG FOR SHABBAT.

It is good to praise Adonai; to sing hymns to Your name, O Most High,
to proclaim Your steadfast love at daybreak, Your faithfulness each night,
with a ten-stringed harp, with voice and lyre together.
You have gladdened me by Your deeds, Adonai; I shout for joy at Your handiwork.
How great are Your works, Adonai, how very subtle Your designs!
A brute cannot know, a fool cannot understand this:

The righteous bloom like a date-palm; they thrive like a cedar in Lebanon;
planted in the house of Adonai, they flourish in the courts of our God.
In old age they still produce fruit; they are full of sap and freshness,
attesting that Adonai is upright, my Rock, in whom there is no wrong.

Psalm 92 — The psalm sung by the Levites in the ancient Temple (*M. Tamid 7:4*), taken by the Rabbis to refer to the longed-for future time when every day will be like Shabbat.

Mechilta d'Rabbi Yishma-eil, Masechta d'Shabta I

Psalms 95-99, 29

L'chah Dodi

Psalms 92-93

Shalom Aleichem

PSALM 93:1-5

יְיָ מָלָךְ גֵּאוּת לָבֵשׁ,
לָבֵשׁ יְיָ עֹז הִתְאַזָּר,
אַף־תִּכּוֹן תֵּבֵל בַּל־תִּמּוֹט.

נָכוֹן כִּסְאֲךָ מֵאָז,
מֵעוֹלָם אָתָּה.

נָשְׂאוּ נְהָרוֹת, יְיָ,
נָשְׂאוּ נְהָרוֹת קוֹלָם,
יִשְׂאוּ נְהָרוֹת דָּכְיָם.
מִקֹּלוֹת מַיִם רַבִּים,
אַדִּירִים מִשְׁבְּרֵי־יָם,
אַדִּיר בַּמָּרוֹם יְיָ.
עֵדֹתֶיךָ נֶאֶמְנוּ מְאֹד,
לְבֵיתְךָ נַאֲוָה־קֹדֶשׁ,
יְיָ לְאֹרֶךְ יָמִים.

ADONAI IS SOVEREIGN,
God is robed in grandeur;
Adonai is robed,
God is girded with strength.
The world stands firm;
it cannot be shaken.
Your throne stands firm from of old;
from eternity You have existed.
The ocean sounds, Adonai,
the ocean sounds its thunder,
the ocean sounds its pounding.
Above the thunder of the mighty waters,
more majestic than the breakers of the sea
is Adonai, majestic on high.
Your decrees are indeed enduring;
holiness befits Your house,
Adonai, for all times.

Psalm 93 — Proclaiming the completion of God's work of creation and the establishment of divine sovereignty over all, this psalm serves as a fitting climax to *Kabbalat Shabbat.* See *M. Tamid* 7:4.

שָׁלוֹם עֲלֵיכֶם, מַלְאֲכֵי הַשָּׁרֵת,
מַלְאֲכֵי עֶלְיוֹן,
מִמֶּלֶךְ מַלְכֵי הַמְּלָכִים,
הַקָּדוֹשׁ בָּרוּךְ הוּא.

בּוֹאֲכֶם לְשָׁלוֹם, מַלְאֲכֵי הַשָּׁלוֹם,
מַלְאֲכֵי עֶלְיוֹן,
מִמֶּלֶךְ מַלְכֵי הַמְּלָכִים,
הַקָּדוֹשׁ בָּרוּךְ הוּא.

בָּרְכוּנִי לְשָׁלוֹם, מַלְאֲכֵי הַשָּׁלוֹם,
מַלְאֲכֵי עֶלְיוֹן,
מִמֶּלֶךְ מַלְכֵי הַמְּלָכִים,
הַקָּדוֹשׁ בָּרוּךְ הוּא.

צֵאתְכֶם לְשָׁלוֹם, מַלְאֲכֵי הַשָּׁלוֹם,
מַלְאֲכֵי עֶלְיוֹן,
מִמֶּלֶךְ מַלְכֵי הַמְּלָכִים,
הַקָּדוֹשׁ בָּרוּךְ הוּא.

PEACE BE TO YOU, O ministering angels, messengers of the Most High,
Majesty of majesties, Holy One of Blessing.

Enter in peace, O messengers of peace, angels of the Most High,
Majesty of majesties, Holy One of Blessing.

Bless me with peace, O messengers of peace, angels of the Most High,
Majesty of majesties, Holy One of Blessing.

Depart in peace, O messengers of peace, angels of the Most High,
Majesty of majesties, Holy One of Blessing.

שָׁלוֹם עֲלֵיכֶם *Shalom Aleichem* — A 17th century Shabbat table-song (זְמִירָה *z'mirah*), probably composed under the influence of Lurianic Kabbalah, alludes to *Shabbat 119b*, which states that two angels accompany each person home from the synagogue as Shabbat begins.

חֲצִי קַדִּישׁ

עַרְבִית לְשַׁבָּת א׳

AR'VIT L'SHABBAT I — SHABBAT EVENING I

יִתְגַּדַּל וְיִתְקַדַּשׁ שְׁמֵהּ רַבָּא
בְּעָלְמָא דִּי בְרָא כִרְעוּתֵהּ,
וְיַמְלִיךְ מַלְכוּתֵהּ
בְּחַיֵּיכוֹן וּבְיוֹמֵיכוֹן
וּבְחַיֵּי דְכָל בֵּית יִשְׂרָאֵל,
בַּעֲגָלָא וּבִזְמַן קָרִיב,
וְאִמְרוּ: אָמֵן.

יְהֵא שְׁמֵהּ רַבָּא מְבָרַךְ
לְעָלַם וּלְעָלְמֵי עָלְמַיָּא.

יִתְבָּרַךְ וְיִשְׁתַּבַּח וְיִתְפָּאַר
וְיִתְרוֹמַם וְיִתְנַשֵּׂא,
וְיִתְהַדַּר וְיִתְעַלֶּה וְיִתְהַלַּל
שְׁמֵהּ דְּקֻדְשָׁא בְּרִיךְ הוּא,
לְעֵלָּא מִן כָּל בִּרְכָתָא וְשִׁירָתָא,
תֻּשְׁבְּחָתָא וְנֶחֱמָתָא,
דַּאֲמִירָן בְּעָלְמָא, וְאִמְרוּ: אָמֵן.

EXALTED and hallowed be God's great name,
in the world which God created, according to plan.
May God's majesty be revealed in the days of our lifetime
and the life of all Israel —
speedily, imminently.
To which we say: Amen.

Blessed be God's great name to all eternity.

Blessed, praised, honored, exalted,
extolled, glorified, adored, and lauded
be the name of the Holy Blessed One,
beyond all earthly words and songs of blessing, praise, and comfort.
To which we say: Amen.

The קַדִּישׁ *Kaddish* is marked by long strings of synonyms of praise. The rhythmic repetition of these words is meant to aid one in achieving a higher meditational state. *Judith Z. Abrams*

Chatzi Kaddish

שְׁמַע וּבִרְכוֹתֶיהָ

SH'MA UVIRCHOTEHA — SH'MA AND ITS BLESSINGS

בָּרְכוּ אֶת יְיָ הַמְבֹרָךְ!
בָּרוּךְ יְיָ הַמְבֹרָךְ
לְעוֹלָם וָעֶד!

PRAISE ADONAI to whom praise is due forever!
Praised be Adonai to whom praise is due,
now and forever!

The *Sh'ma* is one of the prayers one may recite in any language. *M. Sotah 7:1*

For those who choose: The prayer leader at the word בָּרְכוּ *Bar'chu* (the call to worship) bends the knees and bows from the waist, and at יְיָ *Adonai* stands straight. בָּרוּךְ יְיָ *Baruch Adonai* is the communal response, whereupon the community repeats the choreography of the first line.

Bar'chu
Maariv Aravim
Ahavat Olam
Sh'ma
V'ahavta
L'maan tizk'ru
Emet Ve-Emunah
Mi Chamochah
Hashkiveinu
V'shamru

E What ritual begins your school day? How does that moment (or the national anthem at a ball game) bring you — and all your classmates — into common purpose?

בָּרוּךְ אַתָּה, יְיָ
אֱלֹהֵינוּ, מֶלֶךְ הָעוֹלָם,
אֲשֶׁר בִּדְבָרוֹ מַעֲרִיב עֲרָבִים,
בְּחָכְמָה פּוֹתֵחַ שְׁעָרִים,
וּבִתְבוּנָה מְשַׁנֶּה עִתִּים
וּמַחֲלִיף אֶת הַזְּמַנִּים,
וּמְסַדֵּר אֶת הַכּוֹכָבִים
בְּמִשְׁמְרוֹתֵיהֶם בָּרָקִיעַ כִּרְצוֹנוֹ.
בּוֹרֵא יוֹם וָלָיְלָה,
גּוֹלֵל אוֹר מִפְּנֵי חֹשֶׁךְ
וְחֹשֶׁךְ מִפְּנֵי אוֹר,
וּמַעֲבִיר יוֹם וּמֵבִיא לָיְלָה,
וּמַבְדִּיל בֵּין יוֹם וּבֵין לָיְלָה,
יְיָ צְבָאוֹת שְׁמוֹ.
אֵל חַי וְקַיָּם,
תָּמִיד יִמְלֹךְ עָלֵינוּ לְעוֹלָם וָעֶד.
בָּרוּךְ אַתָּה, יְיָ, הַמַּעֲרִיב עֲרָבִים.

BLESSED are You, Adonai our God, Ruler of the universe,
who speaks the evening into being,
skillfully opens the gates,
thoughtfully alters the time and changes the seasons,
and arranges the stars in their heavenly courses according to plan.
You are Creator of day and night,
rolling light away from darkness and darkness from light,
transforming day into night and distinguishing one from the other.
Adonai Tz'vaot is Your Name.
Ever-living God, may You reign continually over us into eternity.
Blessed are You, Adonai, who brings on evening.

בָּרוּךְ אַתָּה, יְיָ, הַמַּעֲרִיב עֲרָבִים.

יְיָ צְבָאוֹת *Adonai Tz'vaot:* this is one of many names that help elucidate God's attributes. God designs, creates and arranges the universe with order and purpose.

The darkness of the first day differed from the darkness that preceded creation. The root of **מַעֲרִיב** *maariv* can also mean "mix," suggesting God mixed light into the primordial darkness. The tiniest spark of light can transform darkness.

Bar'chu
Maariv Aravim
Ahavat Olam
Sh'ma
V'ahavta
L'maan tizk'ru
Emet Ve-Emunah
Mi Chamochah
Hashkiveinu
V'shamru

I Follow the sun's path across the sky. Watch the shadows lengthen as darkness descends. When the sun is gone, imagine that dawn and dusk are racing around the earth to return to where you stand. Does awareness of that ongoing cycle lessen the sense of fear that can result from the loss of light? Does time flow more quickly?

בָּרְכוּ
מַעֲרִיב עֲרָבִים
אַהֲבַת עוֹלָם
שְׁמַע
וְאָהַבְתָּ
לְמַעַן תִּזְכְּרוּ
אֱמֶת וֶאֱמוּנָה
מִי־כָמֹכָה
הַשְׁכִּיבֵנוּ
וְשָׁמְרוּ

אַהֲבַת עוֹלָם

בֵּית יִשְׂרָאֵל עַמְּךָ אָהָבְתָּ,
תּוֹרָה וּמִצְוֹת,
חֻקִּים וּמִשְׁפָּטִים, אוֹתָנוּ לִמַּדְתָּ.
עַל כֵּן, יְיָ אֱלֹהֵינוּ,
בְּשָׁכְבֵנוּ וּבְקוּמֵנוּ
נָשִׂיחַ בְּחֻקֶּיךָ,
וְנִשְׂמַח בְּדִבְרֵי תוֹרָתְךָ
וּבְמִצְוֹתֶיךָ לְעוֹלָם וָעֶד.
כִּי הֵם חַיֵּינוּ וְאֹרֶךְ יָמֵינוּ
וּבָהֶם נֶהְגֶּה יוֹמָם וָלָיְלָה.
וְאַהֲבָתְךָ
אַל תָּסִיר מִמֶּנּוּ לְעוֹלָמִים.
בָּרוּךְ אַתָּה, יְיָ,
אוֹהֵב עַמּוֹ יִשְׂרָאֵל.

EVERLASTING LOVE You offered Your people Israel
by teaching us Torah and mitzvot, laws and precepts.
Therefore, Adonai our God,
when we lie down and when we rise up,
we will meditate on Your laws and Your commandments.
We will rejoice in Your Torah for ever.
Day and night we will reflect on them
for they are our life and doing them lengthens our days.
Never remove Your love from us.
Praise to You, Adonai, who loves Your people Israel.

בָּרוּךְ אַתָּה, יְיָ, אוֹהֵב עַמּוֹ יִשְׂרָאֵל.

שְׁמַע יִשְׂרָאֵל יְהוָה אֱלֹהֵינוּ יְהוָה אֶחָד!

Hear, O Israel, Adonai is our God, Adonai is One!

שְׁמַע יִשְׂרָאֵל *Sh'ma Yisrael . . . Hear, O Israel . . .* Deuteronomy 6:4

בָּרוּךְ שֵׁם כְּבוֹד מַלְכוּתוֹ לְעוֹלָם וָעֶד.

Blessed is God's glorious majesty forever and ever.

The enlarged ע *ayin* at the end of שמע *Sh'ma* (Hear) and the enlarged ד *dalet* at the end of אחד *echad* (one) combine to spell עד *eid* (witness). We recite the *Sh'ma* to bear witness to the Oneness of God.

בָּרוּךְ שֵׁם כְּבוֹד *Baruch shem k'vod . . . Blessed is God's glorious . . . M. Yoma 3:8, inspired by* Nehemiah 9:5

וְאָהַבְתָּ אֵת יְיָ אֱלֹהֶיךָ
בְּכָל־לְבָבְךָ וּבְכָל־נַפְשְׁךָ וּבְכָל־
מְאֹדֶךָ: וְהָיוּ הַדְּבָרִים הָאֵלֶּה
אֲשֶׁר אָנֹכִי מְצַוְּךָ הַיּוֹם עַל־
לְבָבֶךָ: וְשִׁנַּנְתָּם לְבָנֶיךָ וְדִבַּרְתָּ
בָּם בְּשִׁבְתְּךָ בְּבֵיתֶךָ וּבְלֶכְתְּךָ
בַדֶּרֶךְ וּבְשָׁכְבְּךָ וּבְקוּמֶךָ:
וּקְשַׁרְתָּם לְאוֹת עַל־יָדֶךָ וְהָיוּ
לְטֹטָפֹת בֵּין עֵינֶיךָ: וּכְתַבְתָּם
עַל־מְזֻזוֹת בֵּיתֶךָ וּבִשְׁעָרֶיךָ:

לְמַעַן תִּזְכְּרוּ וַעֲשִׂיתֶם אֶת־
כָּל־מִצְוֹתָי וִהְיִיתֶם קְדֹשִׁים
לֵאלֹהֵיכֶם: אֲנִי יְיָ אֱלֹהֵיכֶם
אֲשֶׁר הוֹצֵאתִי אֶתְכֶם מֵאֶרֶץ
מִצְרַיִם לִהְיוֹת לָכֶם לֵאלֹהִים
אֲנִי יְיָ אֱלֹהֵיכֶם:

בָּרְכוּ

מַעֲרִיב עֲרָבִים

אַהֲבַת עוֹלָם

שְׁמַע

וְאָהַבְתָּ

לְמַעַן תִּזְכְּרוּ

אֱמֶת וֶאֱמוּנָה

מִי־כָמֹכָה

הַשְׁכִּיבֵנוּ

וְשָׁמְרוּ

YOU SHALL LOVE Adonai your God with all your heart,
with all your soul, and with all your might.
Take to heart these instructions with which I charge you this day.
Impress them upon your children.
Recite them when you stay at home and when you are away,
when you lie down and when you get up.
Bind them as a sign on your hand and let them serve as a symbol on your forehead;
inscribe them on the doorposts of your house and on your gates.

Thus you shall remember to observe all My commandments
and to be holy to your God.
I am Adonai, your God, who brought you out of the land of Egypt to be your God:
I am Adonai your God.

יְיָ אֱלֹהֵיכֶם אֱמֶת.

For those who choose: At the end of the שְׁמַע *Sh'ma,* after the words יְיָ אֱלֹהֵיכֶם *Adonai Eloheichem,* the word אֱמֶת *emet* ("true") is added as an immediate affirmation of its truth.

וְאָהַבְתָּ *V'ahavta . . . You shall love . . .* Deuteronomy 6:5–9

לְמַעַן תִּזְכְּרוּ *L'maan tizk'ru . . . Thus you shall remember . . .* Numbers 15:40–41

Bar'chu

Maariv Aravim

Ahavat Olam

Sh'ma

V'ahavta

L'maan tizk'ru

Emet Ve-Emunah

Mi Chamochah

Hashkiveinu

V'shamru

בָּרְכוּ

מַעֲרִיב עֲרָבִים

אַהֲבַת עוֹלָם

שְׁמַע

וְאָהַבְתָּ

לְמַעַן תִּזְכְּרוּ

אֱמֶת וֶאֱמוּנָה

מִי־כָמְֽכָה

הַשְׁכִּיבֵנוּ

וְשָׁמְרוּ

אֱמֶת וֶאֱמוּנָה כָּל־זֹאת

וְקַיָּם עָלֵינוּ, כִּי הוּא יְיָ אֱלֹהֵינוּ

וְאֵין זוּלָתוֹ, וַאֲנַֽחְנוּ יִשְׂרָאֵל עַמּוֹ.

הַפּוֹדֵנוּ מִיַּד מְלָכִים, מַלְכֵּנוּ

הַגּוֹאֲלֵנוּ מִכַּף כָּל־הֶעָרִיצִים,

הָעוֹשֶׂה גְדוֹלוֹת עַד אֵין חֵקֶר

וְנִפְלָאוֹת עַד אֵין מִסְפָּר, הַשָּׂם

נַפְשֵׁנוּ בַּחַיִּים, וְלֹא נָתַן לַמּוֹט

רַגְלֵנוּ, הָעֹשֶׂה לָּֽנוּ נִסִּים

בְּפַרְעֹה, אוֹתוֹת וּמוֹפְתִים

בְּאַדְמַת בְּנֵי חָם. וַיּוֹצֵא אֶת־

עַמּוֹ יִשְׂרָאֵל מִתּוֹכָם לְחֵרוּת

עוֹלָם. וְרָאוּ בָנָיו גְּבוּרָתוֹ, שִׁבְּחוּ

וְהוֹדוּ לִשְׁמוֹ. וּמַלְכוּתוֹ בְרָצוֹן

קִבְּלוּ עֲלֵיהֶם. מֹשֶׁה וּמִרְיָם

וּבְנֵי יִשְׂרָאֵל לְךָ עָנוּ שִׁירָה

בְּשִׂמְחָה רַבָּה, וְאָמְרוּ כֻלָּם:

Aʟʟ ᴛʜɪѕ ᴡᴇ ʜᴏʟᴅ to be true and trustworthy for us.
You alone are our God, and we are Israel Your people.
You are our Sovereign and Savior,
who delivers us from oppressors' hands
and saves us from tyrants' fists.
You work wonders without number, marvels beyond count.
You give us life and steady our footsteps.
You performed miracles for us before Pharaoh,
signs and wonders in the land of the Egyptians;
You led Your people Israel out from their midst to freedom for all time.
When Your children witnessed Your dominance
they praised Your Name in gratitude.
And they accepted Your sovereignty —
Moses, Miriam and all Israel sang to You together,
lifting their voices joyously:

הָעֹשֶׂה גְדוֹלוֹת *Haoseh g'dolot . . . You work wonders . . .* Job 9:10

הַשָּׂם נַפְשֵׁנוּ בַּחַיִּים *Hasam nafsheinu bachayim . . . You give us life . . .* Psalm 66:9

מִי־כָמֹֽכָה בָּאֵלִם, יְיָ!
מִי כָּמֹֽכָה נֶאְדָּר בַּקֹּֽדֶשׁ,
נוֹרָא תְהִלֹּת, עֹֽשֵׂה פֶֽלֶא!

מַלְכוּתְךָ רָאוּ בָנֶֽיךָ,
בּוֹקֵֽעַ יָם לִפְנֵי מֹשֶׁה וּמִרְיָם.
זֶה אֵלִי, עָנוּ וְאָמְרוּ,
יְיָ יִמְלֹךְ לְעֹלָם וָעֶד!

וְנֶאֱמַר: כִּי פָדָה יְיָ אֶת־יַעֲקֹב,
וּגְאָלוֹ מִיַּד חָזָק מִמֶּֽנּוּ.
בָּרוּךְ אַתָּה, יְיָ, גָּאַל יִשְׂרָאֵל.

בָּרְכוּ

מַעֲרִיב עֲרָבִים

אַהֲבַת עוֹלָם

שְׁמַע

וְאָהַבְתָּ

לְמַֽעַן תִּזְכְּרוּ

אֱמֶת וֶאֱמוּנָה

מִי־כָמֹֽכָה

הַשְׁכִּיבֵֽנוּ

וְשָׁמְרוּ

WHO IS LIKE YOU, O God,
among the gods that are worshipped?
Who is like You, majestic in holiness,
awesome in splendor, working wonders?

Your children witnessed Your sovereignty,
the sea splitting before Moses and Miriam.
"This is our God!" they cried.
"Adonai will reign forever and ever!"

Thus it is said,
"Adonai redeemed Jacob,
from a hand stronger
than his own."
Praised are You, Adonai, for redeeming Israel.

בָּרוּךְ אַתָּה, יְיָ, גָּאַל יִשְׂרָאֵל.

מִי־כָמֹֽכָה *Mi chamochah . . . Who is like You . . .* Exodus 15:11

זֶה אֵלִי *Zeh Eli . . . This is our God . . .* Exodus 15:2

יְיָ יִמְלֹךְ *Adonai yimloch . . . Adonai will reign . . .* Exodus 15:18

כִּי פָדָה יְיָ *Ki fadah Adonai . . . Adonai redeemed . . .* Jeremiah 31:10

הַשְׁכִּיבֵנוּ, יְיָ אֱלֹהֵינוּ,
לְשָׁלוֹם, וְהַעֲמִידֵנוּ שׁוֹמְרֵנוּ לְחַיִּים,
וּפְרֹשׂ עָלֵינוּ סֻכַּת שְׁלוֹמֶךָ,
וְתַקְּנֵנוּ בְּעֵצָה טוֹבָה מִלְּפָנֶיךָ,
וְהוֹשִׁיעֵנוּ לְמַעַן שְׁמֶךָ.
וְהָגֵן בַּעֲדֵנוּ,
וְהָסֵר מֵעָלֵינוּ אוֹיֵב, דֶּבֶר,
וְחֶרֶב, וְרָעָב, וְיָגוֹן,
וְהַרְחֵק מִמֶּנּוּ עָוֺן וָפֶשַׁע.
וּבְצֵל כְּנָפֶיךָ תַּסְתִּירֵנוּ,
כִּי אֵל שׁוֹמְרֵנוּ וּמַצִּילֵנוּ אָתָּה,
כִּי אֵל חַנּוּן וְרַחוּם אָתָּה.
וּשְׁמֹר צֵאתֵנוּ וּבוֹאֵנוּ
לְחַיִּים וּלְשָׁלוֹם
מֵעַתָּה וְעַד עוֹלָם.
בָּרוּךְ אַתָּה, יְיָ,
הַפּוֹרֵשׂ סֻכַּת שָׁלוֹם עָלֵינוּ
וְעַל כָּל עַמּוֹ יִשְׂרָאֵל וְעַל יְרוּשָׁלָיִם.

GRANT, O GOD, that we lie down in peace,
and raise us up, our Guardian, to life renewed.
Spread over us the shelter of Your peace.
Guide us with Your good counsel; for Your Name's sake, be our help.
Shield and shelter us beneath the shadow of Your wings.
Defend us against enemies, illness, war, famine and sorrow.
Distance us from wrongdoing.
For You, God, watch over us and deliver us. For You, God, are gracious and merciful.
Guard our going and coming, to life and to peace, evermore.

Blessed are You, Adonai, Guardian of Israel, whose shelter of peace is spread over us,
over all Your people Israel, and over Jerusalem.

בָּרוּךְ אַתָּה, יְיָ, הַפּוֹרֵשׂ סֻכַּת שָׁלוֹם עָלֵינוּ
וְעַל כָּל עַמּוֹ יִשְׂרָאֵל וְעַל יְרוּשָׁלָיִם.

Grant, O God, that we lie down in peace . . . Following a reading from *Seder Rav Amram*, our first known comprehensive prayerbook, circa 860 C.E.

Bar'chu
Maariv Aravim
Ahavat Olam
Sh'ma
V'ahavta
L'maan tizk'ru
Emet Ve-Emunah
Mi Chamochah
Hashkiveinu
V'shamru

בָּרְכוּ
מַעֲרִיב עֲרָבִים
אַהֲבַת עוֹלָם
שְׁמַע
וְאָהַבְתָּ
לְמַעַן תִּזְכְּרוּ
אֱמֶת וֶאֱמוּנָה
מִי־כָמֹכָה
הַשְׁכִּיבֵנוּ
וְשָׁמְרוּ

Select either V'shamru or Yism'chu

וְשָׁמְרוּ בְנֵי יִשְׂרָאֵל
אֶת־הַשַּׁבָּת,
לַעֲשׂוֹת אֶת־הַשַּׁבָּת לְדֹרֹתָם
בְּרִית עוֹלָם.
בֵּינִי וּבֵין בְּנֵי יִשְׂרָאֵל
אוֹת הִיא לְעֹלָם,
כִּי־שֵׁשֶׁת יָמִים עָשָׂה יְיָ
אֶת־הַשָּׁמַיִם וְאֶת־הָאָרֶץ,
וּבַיּוֹם הַשְּׁבִיעִי שָׁבַת וַיִּנָּפַשׁ.

THE PEOPLE OF ISRAEL shall keep Shabbat,
observing Shabbat throughout the ages as a covenant for all time.
It is a sign for all time between Me and the people of Israel.
For in six days Adonai made heaven and earth,
and on the seventh day God ceased from work and was refreshed.

יִשְׂמְחוּ בְמַלְכוּתְךָ
שׁוֹמְרֵי שַׁבָּת וְקוֹרְאֵי עֹנֶג.
עַם מְקַדְּשֵׁי שְׁבִיעִי,
כֻּלָּם יִשְׂבְּעוּ וְיִתְעַנְּגוּ מִטּוּבֶךָ.
וְהַשְּׁבִיעִי רָצִיתָ בּוֹ וְקִדַּשְׁתּוֹ,
חֶמְדַּת יָמִים אוֹתוֹ קָרָאתָ,
זֵכֶר לְמַעֲשֵׂה בְרֵאשִׁית.

THOSE WHO KEEP SHABBAT by calling it a delight
will rejoice in Your realm.
The people that hallow Shabbat will delight in Your goodness.
For, being pleased with the Seventh Day, You hallowed it
as the most precious of days, drawing our attention to the work of Creation.

וְשָׁמְרוּ בְנֵי יִשְׂרָאֵל *V'shamru v'nei Yisrael . . . The people of Israel shall keep . . .* Exodus 31:16–17

יִשְׂמְחוּ *Yism'chu* contains twenty-four Hebrew words said to correspond to the twenty-four hours of Shabbat.

אָבוֹת וְאִמָּהוֹת
גְּבוּרוֹת
קְדֻשָּׁה
קְדֻשַּׁת הַיּוֹם
עֲבוֹדָה
הוֹדָאָה
שָׁלוֹם
תְּפִלַּת הַלֵּב

T'FILAH

אֲדֹנָי, שְׂפָתַי תִּפְתָּח,
וּפִי יַגִּיד תְּהִלָּתֶךָ.

ADONAI, open up my lips,
that my mouth may declare Your praise.

For those who choose: Before reciting תְּפִלָּה *T'filah*, one takes three steps forward.

אֲדֹנָי, שְׂפָתַי תִּפְתָּח *Adonai, s'fatai tiftach . . . Adonai, open up my lips . . .* Psalm 51:17

P What if you stopped after this prayer? What if you are not ready, if God has not opened your lips? How do you know if you are ready to pray?

בָּרוּךְ אַתָּה, יְיָ אֱלֹהֵינוּ
וֵאלֹהֵי אֲבוֹתֵינוּ וְאִמּוֹתֵינוּ, אֱלֹהֵי
אַבְרָהָם, אֱלֹהֵי יִצְחָק וֵאלֹהֵי יַעֲקֹב,
אֱלֹהֵי שָׂרָה, אֱלֹהֵי רִבְקָה, אֱלֹהֵי
רָחֵל וֵאלֹהֵי לֵאָה. הָאֵל הַגָּדוֹל
הַגִּבּוֹר וְהַנּוֹרָא, אֵל עֶלְיוֹן, גּוֹמֵל
חֲסָדִים טוֹבִים וְקוֹנֵה הַכֹּל, וְזוֹכֵר
חַסְדֵי אָבוֹת וְאִמָּהוֹת, וּמֵבִיא גְּאֻלָּה
לִבְנֵי בְנֵיהֶם לְמַעַן שְׁמוֹ בְּאַהֲבָה.

*SHABBAT SHUVAH — זָכְרֵנוּ לְחַיִּים,
מֶלֶךְ חָפֵץ בַּחַיִּים,
וְכָתְבֵנוּ בְּסֵפֶר הַחַיִּים,
לְמַעַנְךָ אֱלֹהִים חַיִּים.

מֶלֶךְ עוֹזֵר וּמוֹשִׁיעַ וּמָגֵן.
בָּרוּךְ אַתָּה, יְיָ,
מָגֵן אַבְרָהָם וְעֶזְרַת שָׂרָה.

BLESSED ARE YOU, Adonai our God, God of our fathers and mothers,
God of Abraham, God of Isaac, and God of Jacob,
God of Sarah, God of Rebecca, God of Rachel, and God of Leah,
the great, mighty and awesome God, transcendent God
who bestows loving kindness, creates everything out of love,
remembers the love of our fathers and mothers,
and brings redemption to their children's children for the sake of the Divine Name.

*SHABBAT SHUVAH — Remember us for life, O Sovereign who delights in life,
and inscribe us in the Book of Life, for Your sake, Living God.

Sovereign, Deliverer, Helper and Shield,
Blessed are You, Adonai, Sarah's Helper, Abraham's Shield.

בָּרוּךְ אַתָּה, יְיָ, מָגֵן אַבְרָהָם וְעֶזְרַת שָׂרָה.

---

*SHABBAT SHUVAH: The Shabbat between Rosh HaShanah and Yom Kippur.

For those who choose: At the beginning and end of the blessing, one bends the knees and bows from the waist at the word בָּרוּךְ *Baruch* and stands straight at the word יְיָ *Adonai.*

אָבוֹת וְאִמָּהוֹת *Avot v'imahot . . .* As God has been gracious to our forebears, so may we receive divine favor.

Avot v'Imahot
G'vurot
K'dushah
K'dushat HaYom
Avodah
Hodaah
Shalom
T'filat HaLev

T We are taught in the Torah: "Honor your father and your mother, that you may long endure on the land that Adonai your God is assigning to you" (Exodus 20:12). When you invoke the names of those who have come before you, what is your responsibility to them?

אַתָּה גִּבּוֹר לְעוֹלָם, אֲדֹנָי,
מְחַיֵּה הַכֹּל (מֵתִים) אַתָּה,
רַב לְהוֹשִׁיעַ.

מַשִּׁיב הָרוּחַ — WINTER*
וּמוֹרִיד הַגֶּשֶׁם.
מוֹרִיד הַטָּל. — SUMMER*

מְכַלְכֵּל חַיִּים בְּחֶסֶד,
מְחַיֵּה הַכֹּל (מֵתִים)
בְּרַחֲמִים רַבִּים, סוֹמֵךְ נוֹפְלִים,
וְרוֹפֵא חוֹלִים, וּמַתִּיר אֲסוּרִים,
וּמְקַיֵּם אֱמוּנָתוֹ לִישֵׁנֵי עָפָר.
מִי כָמוֹךָ בַּעַל גְּבוּרוֹת
וּמִי דּוֹמֶה לָּךְ, מֶלֶךְ מֵמִית
וּמְחַיֶּה וּמַצְמִיחַ יְשׁוּעָה.

— SHABBAT SHUVAH
מִי כָמוֹךָ אַב הָרַחֲמִים,
זוֹכֵר יְצוּרָיו לְחַיִּים בְּרַחֲמִים.

וְנֶאֱמָן אַתָּה לְהַחֲיוֹת הַכֹּל (מֵתִים).
בָּרוּךְ אַתָּה, יְיָ, מְחַיֵּה הַכֹּל (הַמֵּתִים).

YOU ARE FOREVER MIGHTY, Adonai; You give life to all (revive the dead).
WINTER — You cause the wind to shift and rain to fall.
SUMMER — You rain dew upon us.
You sustain life through love, giving life to all (reviving the dead) through great compassion, supporting the fallen, healing the sick, freeing the captive, keeping faith with those who sleep in the dust. Who is like You, Source of mighty acts? Who resembles You, a Sovereign who takes and gives life, causing deliverance to spring up and faithfully giving life to all (reviving that which is dead)?
SHABBAT SHUVAH — Who is like You, Compassionate God,
who mercifully remembers Your creatures for life?
Blessed are You, Adonai, who gives life to all (who revives the dead).

בָּרוּךְ אַתָּה, יְיָ, מְחַיֵּה הַכֹּל (הַמֵּתִים).

מַשִּׁיב הָרוּחַ / מוֹרִיד הַטָּל *Mashiv haruach / Morid hatal* — with these words, we join our Israeli brothers and sisters in their prayers for seasonal rains in the Land of Israel.

מוֹרִיד הַטָּל *Morid hatal . . . You rain dew upon us . . .* A seasonal insertion into the *G'vurot* acknowledges God as the Source of the power of nature. The variations in climate like growth and decay, birth and death, are part of the fixed pattern of the universe created by God. In his prayerbook, *Minhag America,* Isaac Mayer Wise used the prayer for dew and rain as a permanent part of this benediction.

*WINTER: *Sh'mini Atzeret / Simchat Torah to Pesach.* SUMMER: *Pesach to Sh'mini Atzeret / Simchat Torah.*

Avot v'Imahot
G'vurot
K'dushah
K'dushat HaYom
Avodah
Hodaah
Shalom
T'filat HaLev

The metaphor "reviving the dead" is widely used rabbinically. The Talmud recommends saying בָּרוּךְ אַתָּה יְיָ, מְחַיֵּה הַמֵּתִים *Baruch atah Adonai, m'chayeih hameitim* for greeting a friend after a lapse of seeing the person for twelve months, and for awakening from sleep.

B'rachot 58b, Y. B'rachot 4:2

אַתָּה קָדוֹשׁ וְשִׁמְךָ קָדוֹשׁ
וּקְדוֹשִׁים בְּכָל יוֹם
יְהַלְלְוּךָ סֶּלָה.*
בָּרוּךְ אַתָּה, יְיָ, הָאֵל הַקָּדוֹשׁ.

*SHABBAT SHUVAH — בָּרוּךְ אַתָּה,
יְיָ, הַמֶּלֶךְ הַקָּדוֹשׁ.

YOU ARE HOLY, Your Name is holy,
and those who are holy praise You every day.*
Blessed are You, Adonai, the Holy God.
בָּרוּךְ אַתָּה, יְיָ, הָאֵל הַקָּדוֹשׁ.

*SHABBAT SHUVAH —
Praised are You, Adonai, Holy Sovereign.
בָּרוּךְ אַתָּה, יְיָ, הַמֶּלֶךְ הַקָּדוֹשׁ.

This third blessing of praise in the עֲמִידָה *Amidah* emphasizes God's holy nature. Even God's name is holy.

Avot v'Imahot
G'vurot
K'dushah
K'dushat HaYom
Avodah
Hodaah
Shalom
T'filat HaLev

E Holiness is difficult to define. If holiness is a quality that we share with God, what in your life inspires you to emulate the Divine, to be holy?

what inspires me to be special is being myself and being kind.

אַתָּה קִדַּשְׁתָּ אֶת יוֹם הַשְּׁבִיעִי לִשְׁמֶךָ,
תַּכְלִית מַעֲשֵׂה שָׁמַיִם וָאָרֶץ,
וּבֵרַכְתּוֹ מִכָּל הַיָּמִים,
וְקִדַּשְׁתּוֹ מִכָּל הַזְּמַנִּים,
וְכֵן כָּתוּב בְּתוֹרָתֶךָ:

וַיְכֻלּוּ הַשָּׁמַיִם וְהָאָרֶץ
וְכָל־צְבָאָם: וַיְכַל אֱלֹהִים בַּיּוֹם
הַשְּׁבִיעִי מְלַאכְתּוֹ אֲשֶׁר עָשָׂה
וַיִּשְׁבֹּת בַּיּוֹם הַשְּׁבִיעִי מִכָּל־
מְלַאכְתּוֹ אֲשֶׁר עָשָׂה: וַיְבָרֶךְ
אֱלֹהִים אֶת־יוֹם הַשְּׁבִיעִי וַיְקַדֵּשׁ אֹתוֹ
כִּי בוֹ שָׁבַת מִכָּל־מְלַאכְתּוֹ
אֲשֶׁר־בָּרָא אֱלֹהִים לַעֲשׂוֹת:

YOU SET ASIDE the seventh day for Your Name, the pinnacle of Creation; and You blessed it above all other days, more sacred than all Festival times. So it is written in Your Torah: The heaven and the earth were finished and all their array. On the seventh day, God had completed the work that had been done, ceasing then on the seventh day from all the work that [God] had done. Then God blessed the seventh day and made it holy, and ceased from all the creative work that God [had chosen] to do.

אֱלֹהֵינוּ וֵאלֹהֵי אֲבוֹתֵינוּ וְאִמּוֹתֵינוּ,
רְצֵה בִמְנוּחָתֵנוּ. קַדְּשֵׁנוּ בְּמִצְוֹתֶיךָ
וְתֵן חֶלְקֵנוּ בְּתוֹרָתֶךָ,
שַׂבְּעֵנוּ מִטּוּבֶךָ וְשַׂמְּחֵנוּ בִּישׁוּעָתֶךָ,
וְטַהֵר לִבֵּנוּ לְעָבְדְּךָ בֶּאֱמֶת,
וְהַנְחִילֵנוּ, יְיָ אֱלֹהֵינוּ,
בְּאַהֲבָה וּבְרָצוֹן שַׁבַּת קָדְשֶׁךָ
וְיָנוּחוּ בָהּ יִשְׂרָאֵל, מְקַדְּשֵׁי שְׁמֶךָ.
בָּרוּךְ אַתָּה, יְיָ, מְקַדֵּשׁ הַשַּׁבָּת.

Our God and God of our ancestors, be pleased with our rest. Sanctify us with Your mitzvot, and grant us a share in Your Torah. Satisfy us with Your goodness and gladden us with Your salvation. Purify our hearts to serve You in truth. In Your gracious love, Adonai our God, grant as our heritage Your Holy Shabbat, that Israel who sanctifies Your Name may rest on it. Praise to You, Adonai, who sanctifies Shabbat.

בָּרוּךְ אַתָּה, יְיָ, מְקַדֵּשׁ הַשַּׁבָּת.

וַיְכֻלּוּ *Vay'chulu . . . The heaven and the earth . . .* Genesis 2:1-3

Avot v'Imahot
G'vurot
K'dushah
K'dushat HaYom
Avodah
Hodaah
Shalom
T'filat HaLev

T Abraham Joshua Heschel writes, "The meaning of Shabbat is to celebrate time rather than space." What do you think he means by that? How can you make Shabbat a celebration of time?

רְצֵה, יְיָ אֱלֹהֵינוּ, בְּעַמְּךָ יִשְׂרָאֵל,
וּתְפִלָּתָם בְּאַהֲבָה תְקַבֵּל,
וּתְהִי לְרָצוֹן תָּמִיד
עֲבוֹדַת יִשְׂרָאֵל עַמֶּךָ.
אֵל קָרוֹב לְכָל קֹרְאָיו,
פְּנֵה אֶל עֲבָדֶיךָ וְחָנֵּנוּ,
שְׁפוֹךְ רוּחֲךָ עָלֵינוּ.

FIND FAVOR, Adonai, our God, with Your people Israel
and accept their prayer in love.
May the worship of Your people Israel always be acceptable.
God who is near to all who call, turn lovingly to Your servants.
Pour out Your spirit upon us.

ROSH CHODESH, PESACH, AND SUKKOT

אֱלֹהֵינוּ וֵאלֹהֵי אֲבוֹתֵינוּ וְאִמּוֹתֵינוּ,
יַעֲלֶה וְיָבֹא וְיִזָּכֵר זִכְרוֹנֵנוּ
וְזִכְרוֹן כָּל עַמְּךָ בֵּית יִשְׂרָאֵל לְפָנֶיךָ,
לְטוֹבָה, לְחֵן וּלְחֶסֶד וּלְרַחֲמִים,
לְחַיִּים וּלְשָׁלוֹם, בְּיוֹם

רֹאשׁ הַחֹדֶשׁ הַזֶּה.
חַג הַמַּצּוֹת הַזֶּה.
חַג הַסֻּכּוֹת הַזֶּה.

זָכְרֵנוּ, יְיָ אֱלֹהֵינוּ, בּוֹ לְטוֹבָה. אָמֵן.
וּפָקְדֵנוּ בוֹ לִבְרָכָה. אָמֵן.
וְהוֹשִׁיעֵנוּ בוֹ לְחַיִּים. אָמֵן.

Our God and God of our fathers and mothers, on this
(first day of the new month) — (day of Pesach) — (day of Sukkot)
be mindful of us and all Your people Israel,
for good, for love, for compassion, life and peace.
Remember us for wellbeing. Amen.
Visit us with blessing. Amen. Help us to a fuller life. Amen.

וְתֶחֱזֶינָה עֵינֵינוּ בְּשׁוּבְךָ
לְצִיּוֹן בְּרַחֲמִים.

LET OUR EYES BEHOLD Your loving return to Zion.
Blessed are You, Adonai, whose Presence returns to Zion.

בָּרוּךְ אַתָּה, יְיָ,
הַמַּחֲזִיר שְׁכִינָתוֹ לְצִיּוֹן.

Avot v'Imahot
G'vurot
K'dushah
K'dushat HaYom
Avodah
Hodaah
Shalom
T'filat HaLev

מוֹדִים אֲנַחְנוּ לָךְ, שָׁאַתָּה הוּא
יְיָ אֱלֹהֵינוּ וֵאלֹהֵי אֲבוֹתֵינוּ וְאִמּוֹתֵינוּ
לְעוֹלָם וָעֶד. צוּר חַיֵּינוּ, מָגֵן יִשְׁעֵנוּ,
אַתָּה הוּא לְדוֹר וָדוֹר.

נוֹדֶה לְּךָ וּנְסַפֵּר תְּהִלָּתֶךָ עַל חַיֵּינוּ
הַמְּסוּרִים בְּיָדֶךָ, וְעַל נִשְׁמוֹתֵינוּ
הַפְּקוּדוֹת לָךְ, וְעַל נִסֶּיךָ שֶׁבְּכָל יוֹם
עִמָּנוּ, וְעַל נִפְלְאוֹתֶיךָ וְטוֹבוֹתֶיךָ
שֶׁבְּכָל עֵת, עֶרֶב וָבֹקֶר וְצָהֳרָיִם.

הַטּוֹב כִּי לֹא כָלוּ רַחֲמֶיךָ, וְהַמְרַחֵם
כִּי לֹא תַמּוּ חֲסָדֶיךָ, מֵעוֹלָם קִוִּינוּ לָךְ.

WE ACKNOWLEDGE with thanks that You are Adonai, our God and the God of our ancestors forever. You are the Rock of our lives, and the Shield of our salvation in every generation. Let us thank You and praise You — for our lives which are in Your hand, for our souls which are in Your care, for Your miracles that we experience every day and for Your wondrous deeds and favors at every time of day: evening, morning and noon. O Good One, whose mercies never end, O Compassionate One, whose kindness never fails, we forever put our hope in You.

וְעַל כֻּלָּם יִתְבָּרַךְ וְיִתְרוֹמַם שִׁמְךָ,
מַלְכֵּנוּ, תָּמִיד לְעוֹלָם וָעֶד.

SHABBAT SHUVAH — וּכְתוֹב לְחַיִּים
טוֹבִים כָּל בְּנֵי בְרִיתֶךָ.

וְכֹל הַחַיִּים יוֹדוּךָ סֶּלָה,
וִיהַלְלוּ אֶת שִׁמְךָ בֶּאֱמֶת,
הָאֵל יְשׁוּעָתֵנוּ וְעֶזְרָתֵנוּ סֶלָה.
בָּרוּךְ אַתָּה, יְיָ,
הַטּוֹב שִׁמְךָ וּלְךָ נָאֶה לְהוֹדוֹת.

For all these things, O Sovereign, let Your Name be forever praised and blessed.

SHABBAT SHUVAH — Inscribe all the children of Your covenant for a good life.

O God, our Redeemer and Helper, let all who live affirm You and praise Your Name in truth. Blessed are You, Adonai, Your Name is Goodness, and You are worthy of thanksgiving.

בָּרוּךְ אַתָּה, יְיָ, הַטּוֹב שִׁמְךָ וּלְךָ נָאֶה לְהוֹדוֹת.

P We can spend an eternity thanking God for all that we have ever received or hope to receive. Does God need to know that we're thankful? Why does our gratitude matter? How does God let us know, "You're welcome"?

For those who choose: On page 176, at the word מוֹדִים *Modim,* one bows at the waist. At יְיָ *Adonai,* one stands up straight.

שָׁלוֹם רָב עַל יִשְׂרָאֵל עַמְּךָ
תָּשִׂים לְעוֹלָם,
כִּי אַתָּה הוּא מֶלֶךְ אָדוֹן
לְכָל הַשָּׁלוֹם.
וְטוֹב בְּעֵינֶיךָ לְבָרֵךְ
אֶת עַמְּךָ יִשְׂרָאֵל
בְּכָל עֵת וּבְכָל שָׁעָה בִּשְׁלוֹמֶךָ.

Shabbat Shuvah — בְּסֵפֶר חַיִּים,
בְּרָכָה וְשָׁלוֹם וּפַרְנָסָה טוֹבָה,
נִזָּכֵר וְנִכָּתֵב לְפָנֶיךָ,
אֲנַחְנוּ וְכָל עַמְּךָ בֵּית יִשְׂרָאֵל,
לְחַיִּים טוֹבִים וּלְשָׁלוֹם.
בָּרוּךְ אַתָּה, יְיָ, עוֹשֵׂה הַשָּׁלוֹם.

בָּרוּךְ אַתָּה, יְיָ,
הַמְבָרֵךְ אֶת עַמּוֹ יִשְׂרָאֵל בַּשָּׁלוֹם.

Grant abundant peace to Israel Your people forever,
for You are the Sovereign God of all peace.
May it be pleasing to You to bless Your people Israel
in every season and moment with Your peace.

Shabbat Shuvah —
In the book of life, blessing, peace and prosperity,
may we be remembered and inscribed by You,
we and all Your people Israel for a good life and for peace.
Blessed are You, Adonai, who makes peace.
בָּרוּךְ אַתָּה, יְיָ, עוֹשֵׂה הַשָּׁלוֹם.

Blessed are You, Adonai, who blesses Your people Israel with peace.
בָּרוּךְ אַתָּה, יְיָ, הַמְבָרֵךְ אֶת עַמּוֹ יִשְׂרָאֵל בַּשָּׁלוֹם.

"Seek peace and pursue it." (Psalm 34:15) . . . The midrash observes, we must "seek" peace *in our own place*, and "pursue it" *in every other place. Numbers Rabbah, Chukat 19:27*

אֱלֹהַי, נְצֹר לְשׁוֹנִי מֵרָע וּשְׂפָתַי
מִדַּבֵּר מִרְמָה, וְלִמְקַלְלַי נַפְשִׁי
תִדּוֹם, וְנַפְשִׁי כֶּעָפָר לַכֹּל תִּהְיֶה.
פְּתַח לִבִּי בְּתוֹרָתֶךָ, וּבְמִצְוֹתֶיךָ
תִּרְדּוֹף נַפְשִׁי. וְכֹל הַחוֹשְׁבִים
עָלַי רָעָה, מְהֵרָה הָפֵר עֲצָתָם
וְקַלְקֵל מַחֲשַׁבְתָּם. עֲשֵׂה לְמַעַן
שְׁמֶךָ, עֲשֵׂה לְמַעַן יְמִינֶךָ, עֲשֵׂה
לְמַעַן קְדֻשָּׁתֶךָ, עֲשֵׂה לְמַעַן
תּוֹרָתֶךָ. לְמַעַן יֵחָלְצוּן יְדִידֶיךָ,
הוֹשִׁיעָה יְמִינְךָ וַעֲנֵנִי.

MY GOD, guard my speech from evil and my lips from deception.
Before those who slander me, I will hold my tongue; I will practice humility.
Open my heart to Your Torah, that I may pursue Your mitzvot.
As for all who think evil of me, cancel their designs and frustrate their schemes.
Act for Your own sake, for the sake of Your Power,
for the sake of Your Holiness, for the sake of Your Torah;
so that Your loved ones may be rescued, save with Your power. And answer me.

יִהְיוּ לְרָצוֹן אִמְרֵי פִי וְהֶגְיוֹן לִבִּי
לְפָנֶיךָ, יְיָ צוּרִי וְגוֹאֲלִי.

May the words of my mouth and the meditations of my heart
be acceptable to You, Adonai, my Rock and my Redeemer.

עֹשֶׂה שָׁלוֹם בִּמְרוֹמָיו,
הוּא יַעֲשֶׂה שָׁלוֹם עָלֵינוּ
וְעַל כָּל יִשְׂרָאֵל, וְעַל כָּל יוֹשְׁבֵי תֵבֵל,
וְאִמְרוּ: אָמֵן.

May the One who makes peace in the high heavens
make peace for us, for all Israel and all who inhabit the earth. Amen.

Reading of the Torah is on page 362. Aleinu and Kaddish begin on page 586.

אֱלֹהַי, נְצֹר לְשׁוֹנִי מֵרָע *Elohai, n'tzor l'shoni meira . . . My God, guard my speech from evil . . .* *based on* Psalm 34:14

לְמַעַן יֵחָלְצוּן *L'maan yeichaltzun . . . so that Your loved ones . . .* Psalm 60:7

יִהְיוּ לְרָצוֹן *Yih'yu l'ratzon . . . May the words of my mouth . . .* Psalm 19:15

Avot v'Imahot
G'vurot
K'dushah
K'dushat HaYom
Avodah
Hodaah
Shalom
T'filat HaLev

T Hannah Senesh's well-known poem "Eli Eli" contains the line, "I pray that these things never end." What are the things you pray will never end?

מָגֵן אָבוֹת וְאִמָּהוֹת

בָּרוּךְ אַתָּה, יְיָ אֱלֹהֵינוּ
וֵאלֹהֵי אֲבוֹתֵינוּ וְאִמּוֹתֵינוּ,
אֱלֹהֵי אַבְרָהָם, אֱלֹהֵי יִצְחָק
וֵאלֹהֵי יַעֲקֹב, אֱלֹהֵי שָׂרָה,
אֱלֹהֵי רִבְקָה, אֱלֹהֵי רָחֵל
וֵאלֹהֵי לֵאָה.
הָאֵל הַגָּדוֹל, הַגִּבּוֹר וְהַנּוֹרָא,
אֵל עֶלְיוֹן, קוֹנֵה שָׁמַיִם וָאָרֶץ.

מָגֵן אָבוֹת וְאִמָּהוֹת בִּדְבָרוֹ,
מְחַיֵּה הַכֹּל (מֵתִים) בְּמַאֲמָרוֹ.
הָאֵל הַקָּדוֹשׁ שֶׁאֵין כָּמוֹהוּ,
הַמֵּנִיחַ לְעַמּוֹ בְּיוֹם שַׁבַּת קָדְשׁוֹ
כִּי בָם רָצָה לְהָנִיחַ לָהֶם
לְפָנָיו נַעֲבֹד בְּיִרְאָה וָפַחַד,
וְנוֹדֶה לִשְׁמוֹ בְּכָל יוֹם תָּמִיד,
מְעוֹן הַבְּרָכוֹת, אֵל הַהוֹדָאוֹת,
אֲדוֹן הַשָּׁלוֹם, מְקַדֵּשׁ הַשַּׁבָּת
וּמְבָרֵךְ שְׁבִיעִי, וּמֵנִיחַ בִּקְדֻשָּׁה לְעַם
מְדֻשְּׁנֵי־עֹנֶג, זֵכֶר לְמַעֲשֵׂה בְרֵאשִׁית.

BLESSED ARE YOU, Adonai our God and God of those who came before us:
God of Abraham, God of Isaac, God of Jacob,
God of Sarah, God of Rebecca, God of Rachel and God of Leah.
Great, mighty and revered God, God transcendent,
Maker of heaven and earth.

With a word, You shielded our ancestors; with a phrase, You give all things life.
Sacred God, beyond compare,
who took joy in providing rest for Your people on Your holy Shabbat day.
We will serve You reverently,
acknowledging on each and every day forever, just who You are:
the One to whom blessing goes, the One to whom thanksgiving is due,
Source of peace, who recalls the time of creation
by sanctifying Shabbat, blessing the seventh day,
and in holiness, granting rest to a people overflowing with joy.

Magein Avot v'Imahot

שַׁחֲרִית לְשַׁבָּת א׳

SHACHARIT L'SHABBAT I — SHABBAT MORNING I

INSPIRATION FOR PRAYER

EACH OF US enters this sanctuary with a different need.

Some hearts are full of peace and gratitude,
overflowing with love and joy.
They are eager to confront the day, to make the world a better place.
They are recovering from illness, or have escaped misfortune.
We rejoice with them.

Some hearts ache with sorrow;
disappointments weigh heavily on them.
Families have been broken; loved ones lie on a bed of pain;
death has taken a cherished loved one.
May our presence and caring bring them comfort.

Some hearts are embittered:
ideals are betrayed and mocked, answers sought in vain,
life has lost its meaning and value.
May the knowledge that we, too, are searching
restore our hope, and renew our faith.

WHO DOES NOT HUNGER:
for friendship, understanding,
warmth, and love.

חֲזַק, חֲזַק וְנִתְחַזֵּק
Let us lend strength to one another
and pray for the welfare of this community.

בִּרְכוֹת הַשַּׁחַר

BIRCHOT HASHACHAR — MORNING BLESSINGS

מוֹדֶה / מוֹדָה אֲנִי לְפָנֶיךָ,
מֶלֶךְ חַי וְקַיָּם,
שֶׁהֶחֱזַרְתָּ בִּי נִשְׁמָתִי בְּחֶמְלָה,
רַבָּה אֱמוּנָתֶךָ.

I OFFER THANKS to You,
ever-living Sovereign,
that You have restored my soul to me in mercy:
How great is Your trust.

P The Rabbis taught that sleep is a sixtieth part of death (*B'rachot* 57b). What does that idea mean to you? How does this relate to the idea of being reborn each day? If you were to live as if you were reborn each day, how would your life look different?

MUSIC SELECTIONS

HINEIH MAH TOV

הִנֵּה מַה טּוֹב וּמַה נָּעִים
שֶׁבֶת אַחִים גַּם יָחַד/
שֶׁבֶת אֲחָיוֹת גַּם יָחַד.

How good and how pleasant it is that brothers/sisters dwell together.

(Psalm 133:1)

PITCHU LI

פִּתְחוּ־לִי שַׁעֲרֵי־צֶדֶק,
אָבֹא־בָם אוֹדֶה יָהּ.

Open the gates of victory for me that I may enter them and praise Adonai.

(Psalm 118:19)

V'TAHEIR LIBEINU

וְטַהֵר לִבֵּנוּ לְעָבְדְּךָ בֶּאֱמֶת.

Purify our hearts to serve You in truth.

KOL HAN'SHAMAH T'HALEIL YAH

כֹּל הַנְּשָׁמָה תְּהַלֵּל יָהּ, הַלְלוּ־יָהּ!

Let all that breathes praise God, Hallelujah! *(Psalm 150:6)*

ESA EINAI

אֶשָּׂא עֵינַי אֶל הֶהָרִים, מֵאַיִן יָבֹא עֶזְרִי?
עֶזְרִי מֵעִם יְיָ, עֹשֵׂה שָׁמַיִם וָאָרֶץ.

I turn my eyes to the mountains; from where will my help come?
My help comes from God, Maker of heaven and earth.

(Psalm 121:1–2)

MAH GADLU

מַה גָּדְלוּ מַעֲשֶׂיךָ יָהּ,
מְאֹד עָמְקוּ מַחְשְׁבֹתֶיךָ. הַלְלוּ־יָהּ.

How great are Your works, Adonai. How very profound Your designs! Hallelujah.
(Psalm 92:6)

MI HA-ISH

מִי הָאִישׁ הֶחָפֵץ חַיִּים,
אֹהֵב יָמִים לִרְאוֹת טוֹב?
נְצֹר לְשׁוֹנְךָ מֵרָע,
וּשְׂפָתֶיךָ מִדַּבֵּר מִרְמָה;
סוּר מֵרָע וַעֲשֵׂה טוֹב,
בַּקֵּשׁ שָׁלוֹם וְרָדְפֵהוּ.

Who is the one who is eager for life, who desires years of good fortune?
Guard your tongue from evil, your lips from deceitful speech.
Shun evil and do good; seek integrity and pursue it.
(Psalm 34:13–15)

HAL'LI

הַלְלִי נַפְשִׁי אֶת־יְיָ.
אֲהַלְלָה יְיָ בְּחַיָּי,
אֲזַמְּרָה לֵאלֹהַי בְּעוֹדִי.

Praise Adonai, O my soul! I will praise Adonai all my life,
sing hymns to my God while I exist. *(Psalm 146:1–2)*

VAANACHNU N'VAREICH YAH

וַאֲנַחְנוּ נְבָרֵךְ יָהּ,
מֵעַתָּה וְעַד עוֹלָם, הַלְלוּיָהּ.

But we will bless Adonai now and forever. Hallelujah. *(Psalm 115:18)*

בְּרוּכִים הַבָּאִים
מוֹדֶה / מוֹדָה אֲנִי
שִׁירֵי שַׁבָּת
צִיצִת
מַה־טֹּבוּ
אֲשֶׁר יָצַר
אֱלֹהַי נְשָׁמָה
נִסִּים בְּכָל יוֹם
לַעֲסוֹק
וְהַעֲרֶב־נָא
אֵלּוּ דְבָרִים
קַדִּישׁ דְּרַבָּנָן

FOR THOSE WHO WEAR TALLIT

AS I WRAP myself in the tallit,
I fulfill the mitzvah of my Creator.

Before putting on tallit

בָּרְכִי נַפְשִׁי אֶת יְיָ!
יְיָ אֱלֹהַי, גָּדַלְתָּ מְּאֹד,
הוֹד וְהָדָר לָבָֽשְׁתָּ.
עֹֽטֶה אוֹר כַּשַּׂלְמָה,
נוֹטֶה שָׁמַֽיִם כַּיְרִיעָה.

BLESS, ADONAI, O my soul!
Adonai my God, how great You are.
You are robed in glory and majesty,
wrapping Yourself in light as in a garment,
spreading forth the heavens like a curtain.

בָּרוּךְ אַתָּה, יְיָ
אֱלֹהֵֽינוּ, מֶֽלֶךְ הָעוֹלָם,
אֲשֶׁר קִדְּשָֽׁנוּ בְּמִצְוֹתָיו
וְצִוָּֽנוּ לְהִתְעַטֵּף בַּצִּיצִת.

BLESSED ARE YOU, Adonai our God,
Sovereign of the universe,
who hallows us with mitzvot,
commanding us to wrap ourselves in the fringes.

וְצִוָּֽנוּ לְהִתְעַטֵּף בַּצִּיצִת *v'tzivanu l'hitateif batzitzit... commanding us to wrap ourselves in the fringes.* This *mitzvah* is drawn from Numbers 15:38–39.

Welcome

Modeh / Modah Ani

Shabbat Songs

Tzitzit

Mah Tovu

Asher Yatzar

Elohai N'shamah

Nisim B'chol Yom

Laasok

V'haarev Na

Eilu D'varim

Kaddish D'Rabanan

מַה־טֹּֽבוּ אֹהָלֶֽיךָ, יַעֲקֹב,
מִשְׁכְּנֹתֶֽיךָ, יִשְׂרָאֵל!

וַאֲנִי בְּרֹב חַסְדְּךָ
אָבוֹא בֵיתֶֽךָ,
אֶשְׁתַּחֲוֶה אֶל־הֵיכַל קָדְשְׁךָ
בְּיִרְאָתֶֽךָ.

יְיָ, אָהַֽבְתִּי מְעוֹן בֵּיתֶֽךָ
וּמְקוֹם מִשְׁכַּן כְּבוֹדֶֽךָ.

וַאֲנִי אֶשְׁתַּחֲוֶה וְאֶכְרָֽעָה,
אֶבְרְכָה לִפְנֵי־יְיָ עֹשִׂי.

וַאֲנִי תְפִלָּתִי־לְךָ, יְיָ,
עֵת רָצוֹן.
אֱלֹהִים, בְּרָב־חַסְדֶּֽךָ,
עֲנֵֽנִי בֶּאֱמֶת יִשְׁעֶֽךָ.

בְּרוּכִים הַבָּאִים

מוֹדֶה / מוֹדָה אֲנִי

שִׁירֵי שַׁבָּת

צִיצִת

מַה־טֹּֽבוּ

אֲשֶׁר יָצַר

אֱלֹהַי נְשָׁמָה

נִסִּים בְּכָל יוֹם

לַעֲסוֹק

וְהַעֲרֶב־נָא

אֵלּוּ דְבָרִים

קַדִּישׁ דְּרַבָּנָן

HOW FAIR are your tents, O Jacob,
your dwellings, O Israel.

I, through Your abundant love, enter Your house;
I bow down in awe at Your holy temple.

Adonai, I love Your temple abode,
the dwelling-place of Your glory.

I will humbly bow down low before Adonai, my Maker.

As for me, may my prayer come to You, Adonai, at a favorable time.
O God, in Your abundant faithfulness, answer me with Your sure deliverance.

The opening words of this passage are from Numbers 24:5 where they are recited by Balaam, the foreign prophet who was commissioned to curse the children of Israel. When he opened his mouth, blessings emerged instead of curses.

וַאֲנִי תְפִלָּתִי *Vaani t'filati . . . As for me, may my prayer . . .* The Hebrew text has often been creatively rendered as "I am my prayer" — All I have to offer in prayer is myself; my prayer begins in humility. *Arthur Green*

מַה־טֹּֽבוּ *Mah tovu . . . How fair . . .* Numbers 24:5

וַאֲנִי בְּרֹב חַסְדְּךָ *Vaani b'rov chasd'cha . . . I, through Your abundant love . . .* Psalm 5:8

יְיָ, אָהַֽבְתִּי *Adonai, ahavti . . . Adonai, I love . . .* Psalm 26:8

וַאֲנִי תְפִלָּתִי *Vaani t'filati . . . As for me, may my prayer . . .* Psalm 69:14

T *Mikdash M'at,* a traditional manual for prayer, teaches that when you see a synagogue from a distance, you should say the opening words of *Mah Tovu.* It also teaches that when you arrive at the door of the synagogue, you should arrange your clothes properly and say, "I, through Your abundant love, enter Your house." Why do you think these rules were created? What other actions might you add to this prayer?

בָּרוּךְ אַתָּה, יְיָ
אֱלֹהֵינוּ, מֶלֶךְ הָעוֹלָם,
אֲשֶׁר יָצַר אֶת הָאָדָם בְּחָכְמָה
וּבָרָא בוֹ נְקָבִים נְקָבִים,
חֲלוּלִים חֲלוּלִים.
גָּלוּי וְיָדוּעַ לִפְנֵי כִסֵּא כְבוֹדֶךָ
שֶׁאִם יִפָּתֵחַ אֶחָד מֵהֶם
אוֹ יִסָּתֵם אֶחָד מֵהֶם,
אִי אֶפְשָׁר לְהִתְקַיֵּם
וְלַעֲמֹד לְפָנֶיךָ.
בָּרוּךְ אַתָּה, יְיָ,
רוֹפֵא כָל בָּשָׂר וּמַפְלִיא לַעֲשׂוֹת.

PRAISE TO YOU, Adonai our God,
Sovereign of the universe,
who formed the human body with skill,
creating the body's many pathways and openings.
It is well known before Your throne of glory
that if one of them be wrongly opened or closed,
it would be impossible to endure and stand before You.
Blessed are You, Adonai, who heals all flesh, working wondrously.

בָּרוּךְ אַתָּה, יְיָ, רוֹפֵא כָל בָּשָׂר וּמַפְלִיא לַעֲשׂוֹת.

E When we are well, we take our bodies for granted and use them to do what we wish. When we are ill and limited by what our bodies can and can't do, we no longer take our health and bodies for granted. Of course, illness isn't the only time we are limited by the abilities of our bodies. Describe a time when your soul sought to fly and the limits of your body pulled you back to earth.

בְּרוּכִים הַבָּאִים
מוֹדֶה / מוֹדָה אֲנִי
שִׁירֵי שַׁבָּת
צִיצִת
מַה־טֹּבוּ
אֲשֶׁר יָצַר
אֱלֹהַי נְשָׁמָה
נִסִּים בְּכָל יוֹם
לַעֲסוֹק
וְהַעֲרֶב־נָא
אֵלּוּ דְבָרִים
קַדִּישׁ דְּרַבָּנָן

אֱלֹהַי, נְשָׁמָה שֶׁנָּתַתָּ בִּי
טְהוֹרָה הִיא.
אַתָּה בְרָאתָהּ, אַתָּה יְצַרְתָּהּ,
אַתָּה נְפַחְתָּהּ בִּי,
וְאַתָּה מְשַׁמְּרָהּ בְּקִרְבִּי.
כָּל זְמַן שֶׁהַנְּשָׁמָה בְּקִרְבִּי,
מוֹדֶה / מוֹדָה אֲנִי לְפָנֶיךָ,
יְיָ אֱלֹהַי
וֵאלֹהֵי אֲבוֹתַי וְאִמּוֹתַי,
רִבּוֹן כָּל הַמַּעֲשִׂים,
אֲדוֹן כָּל הַנְּשָׁמוֹת.
בָּרוּךְ אַתָּה, יְיָ,
אֲשֶׁר בְּיָדוֹ נֶפֶשׁ כָּל חַי
וְרוּחַ כָּל בְּשַׂר אִישׁ.

MY GOD, the soul You have given me is pure.
You created it, You shaped it, You breathed it into me,
and You protect it within me.
For as long as my soul is within me,
I offer thanks to You,
Adonai, my God
and God of my ancestors,
Source of all Creation, Sovereign of all souls.
Praised are You, Adonai,
in whose hand is every living soul and the breath of humankind.

בָּרוּךְ אַתָּה, יְיָ, אֲשֶׁר בְּיָדוֹ נֶפֶשׁ כָּל חַי וְרוּחַ כָּל בְּשַׂר אִישׁ.

אֱלֹהַי, נְשָׁמָה *Elohai, n'shamah . . . My God, the soul . . . based on B'rachot 60b*

אֲשֶׁר בְּיָדוֹ *asher b'yado . . . in whose hand . . .* Job 12:10

נִסִּים בְּכָל יוֹם
NISIM B'CHOL YOM — FOR DAILY MIRACLES

For awakening

בָּרוּךְ אַתָּה, יְיָ
אֱלֹהֵינוּ, מֶלֶךְ הָעוֹלָם,
אֲשֶׁר נָתַן לַשֶּׂכְוִי בִינָה
לְהַבְחִין בֵּין יוֹם וּבֵין לָיְלָה.

PRAISE TO YOU, Adonai our God, Sovereign of the universe,
who has given the mind the ability to distinguish day from night.

For vision

בָּרוּךְ אַתָּה, יְיָ
אֱלֹהֵינוּ, מֶלֶךְ הָעוֹלָם,
פּוֹקֵחַ עִוְרִים.

PRAISE TO YOU, Adonai our God, Sovereign of the universe,
who opens the eyes of the blind.

For the ability to stretch

בָּרוּךְ אַתָּה, יְיָ
אֱלֹהֵינוּ, מֶלֶךְ הָעוֹלָם,
מַתִּיר אֲסוּרִים.

PRAISE TO YOU, Adonai our God, Sovereign of the universe,
who frees the captive.

For rising to the new day

בָּרוּךְ אַתָּה, יְיָ
אֱלֹהֵינוּ, מֶלֶךְ הָעוֹלָם,
זוֹקֵף כְּפוּפִים.

PRAISE TO YOU, Adonai our God, Sovereign of the universe,
who lifts up the fallen.

נִסִּים בְּכָל יוֹם *Nisim b'chol yom . . . For daily miracles . . .* These morning blessings evoke wonder at awakening to physical life: we open our eyes, clothe our bodies, and walk again with purpose; in spiritual life also, we are created in God's image, are free human beings, and as Jews, celebrate the joy and destiny of our people, Israel.

Though they are intended literally, we may perceive each blessing spiritually.

For Daily Miracles — Inspiration for blessings three to five comes from Psalm 146:7–8.

P We thank God for clothing the naked, redeeming the oppressed, even פּוֹקֵחַ עִוְרִים *pokei-ach ivrim* — opening the eyes of the blind. What does the prayer mean by attributing these acts to God? What is our responsibility in these matters?

נִסִּים בְּכָל יוֹם

NISIM B'CHOL YOM — FOR DAILY MIRACLES

For firm earth to stand upon

בָּרוּךְ אַתָּה, יְיָ
אֱלֹהֵינוּ, מֶלֶךְ הָעוֹלָם,
רוֹקַע הָאָרֶץ עַל הַמָּיִם.

PRAISE TO YOU, Adonai our God, Sovereign of the universe,
who stretches the earth over the waters.

For the gift of motion

בָּרוּךְ אַתָּה, יְיָ
אֱלֹהֵינוּ, מֶלֶךְ הָעוֹלָם,
הַמֵּכִין מִצְעֲדֵי גָבֶר.

PRAISE TO YOU, Adonai our God, Sovereign of the universe,
who strengthens our steps.

For clothing the body

בָּרוּךְ אַתָּה, יְיָ
אֱלֹהֵינוּ, מֶלֶךְ הָעוֹלָם,
מַלְבִּישׁ עֲרֻמִּים.

PRAISE TO YOU, Adonai our God, Sovereign of the universe,
who clothes the naked.

For renewed enthusiasm for life

בָּרוּךְ אַתָּה, יְיָ
אֱלֹהֵינוּ, מֶלֶךְ הָעוֹלָם,
הַנּוֹתֵן לַיָּעֵף כֹּחַ.

PRAISE TO YOU, Adonai our God, Sovereign of the universe,
who gives strength to the weary.

For reawakening

בָּרוּךְ אַתָּה, יְיָ
אֱלֹהֵינוּ, מֶלֶךְ הָעוֹלָם,
הַמַּעֲבִיר שֵׁנָה מֵעֵינָי,
וּתְנוּמָה מֵעַפְעַפָּי.

PRAISE TO YOU, Adonai our God, Sovereign of the universe,
who removes sleep from the eyes, slumber from the eyelids.

Welcome
Modeh / Modah Ani
Shabbat Songs
Tzitzit
Mah Tovu
Asher Yatzar
Elohai N'shamah
Nisim B'chol Yom
Laasok
V'haarev Na
Eilu D'varim
Kaddish D'Rabanan

נִסִּים בְּכָל יוֹם

NISIM B'CHOL YOM — FOR DAILY MIRACLES

For being in the image of God

בָּרוּךְ אַתָּה, יְיָ
אֱלֹהֵינוּ, מֶלֶךְ הָעוֹלָם,
שֶׁעָשַׂנִי בְּצֶלֶם אֱלֹהִים.

PRAISE TO YOU, Adonai our God, Sovereign of the universe,
who made me in the image of God.

For being a free person

בָּרוּךְ אַתָּה, יְיָ
אֱלֹהֵינוּ, מֶלֶךְ הָעוֹלָם,
שֶׁעָשַׂנִי בֶּן / בַּת חוֹרִין.

PRAISE TO YOU, Adonai our God, Sovereign of the universe,
who has made me free.

For being a Jew

בָּרוּךְ אַתָּה, יְיָ
אֱלֹהֵינוּ, מֶלֶךְ הָעוֹלָם,
שֶׁעָשַׂנִי יִשְׂרָאֵל.

PRAISE TO YOU, Adonai our God, Sovereign of the universe,
who has made me a Jew.

For purpose

בָּרוּךְ אַתָּה, יְיָ
אֱלֹהֵינוּ, מֶלֶךְ הָעוֹלָם,
אוֹזֵר יִשְׂרָאֵל בִּגְבוּרָה.

PRAISE TO YOU, Adonai our God, Sovereign of the universe,
who girds Israel with strength.

For harmony

בָּרוּךְ אַתָּה, יְיָ
אֱלֹהֵינוּ, מֶלֶךְ הָעוֹלָם,
עוֹטֵר יִשְׂרָאֵל בְּתִפְאָרָה.

PRAISE TO YOU, Adonai our God, Sovereign of the universe,
who crowns Israel with splendor.

E When have you felt that your Jewish heritage has אוֹזֵר בִּגְבוּרָה *ozeir big'vurah* — girded you with strength? When have your actions עוֹטֵר יִשְׂרָאֵל בְּתִפְאָרָה *oteir Yisrael b'tifarah* — crowned Israel with splendor?

בָּרוּךְ אַתָּה, יְיָ
אֱלֹהֵינוּ, מֶלֶךְ הָעוֹלָם,
אֲשֶׁר קִדְּשָׁנוּ בְּמִצְוֹתָיו
וְצִוָּנוּ לַעֲסוֹק בְּדִבְרֵי תוֹרָה.

BLESSED ARE YOU, Adonai our God,
Sovereign of the universe,
who hallows us with mitzvot,
commanding us to engage with words of Torah.

וְהַעֲרֶב־נָא יְיָ אֱלֹהֵינוּ
אֶת־דִּבְרֵי תוֹרָתְךָ בְּפִינוּ,
וּבְפִי עַמְּךָ בֵּית יִשְׂרָאֵל,
וְנִהְיֶה אֲנַחְנוּ וְצֶאֱצָאֵינוּ,
וְצֶאֱצָאֵי עַמְּךָ בֵּית יִשְׂרָאֵל,
כֻּלָּנוּ יוֹדְעֵי שְׁמֶךָ,
וְלוֹמְדֵי תוֹרָתֶךָ לִשְׁמָהּ.
בָּרוּךְ אַתָּה יְיָ,
הַמְלַמֵּד תּוֹרָה לְעַמּוֹ יִשְׂרָאֵל.

O ADONAI, our God,
let the words of Torah be sweet in our mouths
and the mouths of Your people Israel,
so that we, our descendants and the descendants of all Your people Israel
may know You, by studying Your Torah for its own sake.
Blessed are You, Adonai, who teaches Torah to Your people Israel.

בָּרוּךְ אַתָּה, יְיָ, הַמְלַמֵּד תּוֹרָה לְעַמּוֹ יִשְׂרָאֵל.

"Descendants" includes men and women who embrace the Jewish people and faith.

The two blessings above (derived from *B'rachot 11b*) are both Torah blessings and are intended to introduce a moment of Torah study, which could be done at this time.

לַעֲסוֹק *Laasok . . . to engage . . .* אֵלּוּ דְבָרִים *Eilu d'varim . . . These are things . . .* The traditional placement of these prayers differs. Here they are linked to emphasize the study of Torah and its influence on our daily ethical behavior.

אֵלּוּ דְבָרִים שֶׁאֵין לָהֶם שִׁעוּר,
שֶׁאָדָם אוֹכֵל פֵּרוֹתֵיהֶם
בָּעוֹלָם הַזֶּה
וְהַקֶּרֶן קַיֶּמֶת לוֹ לָעוֹלָם הַבָּא.
וְאֵלּוּ הֵן:
כִּבּוּד אָב וָאֵם,
וּגְמִילוּת חֲסָדִים,
וְהַשְׁכָּמַת בֵּית הַמִּדְרָשׁ
שַׁחֲרִית וְעַרְבִית,
וְהַכְנָסַת אוֹרְחִים,
וּבִקּוּר חוֹלִים,
וְהַכְנָסַת כַּלָּה,
וּלְוָיַת הַמֵּת,
וְעִיּוּן תְּפִלָּה,
וַהֲבָאַת שָׁלוֹם בֵּין אָדָם לַחֲבֵרוֹ.
וְתַלְמוּד תּוֹרָה כְּנֶגֶד כֻּלָּם.

THESE ARE THINGS that are limitless,
of which a person enjoys the fruit of the world,
while the principal remains in the world to come.
They are: honoring one's father and mother,
engaging in deeds of compassion,
arriving early for study, morning and evening,
dealing graciously with guests, visiting the sick,
providing for the wedding couple,
accompanying the dead for burial,
being devoted in prayer,
and making peace among people.
But the study of Torah encompasses them all.

כִּבּוּד אָב וָאֵם *Kibud av va-eim — honoring one's father and mother.* What can it mean to honor? Sometimes honoring one's parent is not easy. The word כִּבּוּד *kibud, honor,* has as its root כָּבֵד *kaveid — heavy;* it can be a burden to honor another. In honoring those who have given us life or sustenance, we honor the Source of Life. *Elyse D. Frishman*

וְהַשְׁכָּמַת בֵּית הַמִּדְרָשׁ *V'hashkamat beit hamidrash . . . arriving early for study . . .* The Rabbis understood this to convey enthusiasm and earnestness. *Yoel Kahn*

תַּלְמוּד תּוֹרָה *Talmud Torah, the study of Torah* offers the knowledge of what is right and how to live justly. Jewish study includes the expectation that the lessons will be applied to life.

אֵלּוּ דְבָרִים *Eilu d'varim . . . These are things . . . based on Peah 1:1*

שֶׁאָדָם אוֹכֵל *She-adam ocheil . . . of which a person enjoys . . . Shabbat 127a*

E Think of someone who has helped you both materially and spiritually when you were in need. Did you thank them at the time? Do you carry a feeling of gratitude toward them that goes beyond each instance of assistance? What would you say to them now if they were with you?

יִתְגַּדַּל וְיִתְקַדַּשׁ שְׁמֵהּ רַבָּא

בְּעָלְמָא דִּי בְרָא כִרְעוּתֵהּ,

וְיַמְלִיךְ מַלְכוּתֵהּ

בְּחַיֵּיכוֹן וּבְיוֹמֵיכוֹן

וּבְחַיֵּי דְכָל בֵּית יִשְׂרָאֵל,

בַּעֲגָלָא וּבִזְמַן קָרִיב,

וְאִמְרוּ אָמֵן.

יְהֵא שְׁמֵהּ רַבָּא מְבָרַךְ

לְעָלַם וּלְעָלְמֵי עָלְמַיָּא.

יִתְבָּרַךְ וְיִשְׁתַּבַּח וְיִתְפָּאַר

וְיִתְרוֹמַם וְיִתְנַשֵּׂא,

וְיִתְהַדָּר וְיִתְעַלֶּה וְיִתְהַלָּל

שְׁמֵהּ דְּקֻדְשָׁא בְּרִיךְ הוּא,

לְעֵלָּא מִן כָּל בִּרְכָתָא וְשִׁירָתָא,

תֻּשְׁבְּחָתָא וְנֶחֱמָתָא,

דַּאֲמִירָן בְּעָלְמָא, וְאִמְרוּ אָמֵן.

בְּרוּכִים הַבָּאִים

מוֹדֶה / מוֹדָה אֲנִי

שִׁירֵי שַׁבָּת

צִיצִת

מַה־טֹּבוּ

אֲשֶׁר יָצַר

אֱלֹהַי נְשָׁמָה

נִסִּים בְּכָל יוֹם

לַעֲסוֹק

וְהַעֲרֶב־נָא

אֵלּוּ דְבָרִים

קַדִּישׁ דְּרַבָּנָן

EXALTED and hallowed be God's great name,
in the world which God created, according to plan.
May God's majesty be revealed in the days of our lifetime
and the life of all Israel —
speedily, imminently.
To which we say: Amen.

Blessed be God's great name to all eternity.

Blessed, praised, honored, exalted,
extolled, glorified, adored, and lauded
be the name of the Holy Blessed One,
beyond all earthly words and songs of blessing, praise, and comfort.
To which we say: Amen.

Kaddish D'Rabanan continues on page 210.

Current scholarship sees the קַדִּישׁ דְּרַבָּנָן *Kaddish d'Rabanan* as but one of many alternative early versions of the *Kaddish*. It emerged in an oral form in the first or second century. Like other forms of the *Kaddish*, it is an elaborate praise of God, calling for the coming of God's ultimate dominion. Its unique name, קַדִּישׁ דְּרַבָּנָן *Kaddish d'Rabanan* ("Kaddish of the Rabbis"), reflects its central paragraph, acknowledging those who study Torah, indicating the role of Torah study as both an intellectual and a spiritual activity. This Kaddish concludes text study that takes place during worship and other occasions.

עַל יִשְׂרָאֵל וְעַל רַבָּנָן,
וְעַל תַּלְמִידֵיהוֹן
וְעַל כָּל תַּלְמִידֵי תַלְמִידֵיהוֹן,
וְעַל כָּל מָן דְּעָסְקִין בְּאוֹרַיְתָא,
דִּי בְּאַתְרָא הָדֵין וְדִי
בְּכָל אֲתַר וַאֲתַר.

יְהֵא לְהוֹן וּלְכוֹן שְׁלָמָא רַבָּא,
חִנָּא וְחִסְדָּא וְרַחֲמִין,
וְחַיִּין אֲרִיכִין,
וּמְזוֹנֵי רְוִיחֵי, וּפֻרְקָנָא,
מִן קֳדָם אֲבוּהוֹן
דִּי בִשְׁמַיָּא וְאַרְעָא וְאִמְרוּ: אָמֵן.

יְהֵא שְׁלָמָא רַבָּא מִן שְׁמַיָּא,
וְחַיִּים טוֹבִים עָלֵינוּ וְעַל כָּל
יִשְׂרָאֵל וְאִמְרוּ: אָמֵן.

עֹשֶׂה שָׁלוֹם בִּמְרוֹמָיו
הוּא בְּרַחֲמָיו יַעֲשֶׂה שָׁלוֹם
עָלֵינוּ וְעַל כָּל יִשְׂרָאֵל, וְעַל כָּל
יוֹשְׁבֵי תֵבֵל, וְאִמְרוּ: אָמֵן.

GOD OF HEAVEN AND EARTH, grant abundant peace
to our people Israel and their rabbis, to our teachers and their disciples,
and to all who engage in the study of Torah here and everywhere.

Let there be for them and for us all, grace, love, and compassion,
a full life, ample sustenance, and salvation from God, and let us respond: Amen.

For us and all Israel, may the blessing of peace and the promise of life come true,
and let us respond: Amen.

May the One who causes peace to reign in the high heavens,
let peace descend on us, and on all Israel, and on all the world,
and let us respond: Amen.

T Jewish tradition teaches עֲשֵׂה לְךָ רַב *Aseih l'cha rav,* "Find yourself a teacher" (*Pirkei Avot* 1:6). Teachers come from many places in our lives. Some, like schoolteachers or coaches, are found in expected places. But sometimes a teacher can be found in a sibling, a friend, or someone you have never met before. Who has been a true teacher for you? How did you find that person? What was the important lesson you learned?

פְּסוּקֵי דְזִמְרָה

P'SUKEI D'ZIMRAH — VERSES OF PRAISE

בָּרוּךְ שֶׁאָמַר וְהָיָה הָעוֹלָם,
בָּרוּךְ הוּא.
בָּרוּךְ עוֹשֶׂה בְרֵאשִׁית,
בָּרוּךְ אוֹמֵר וְעוֹשֶׂה,
בָּרוּךְ גּוֹזֵר וּמְקַיֵּם,
בָּרוּךְ מְרַחֵם עַל הָאָרֶץ,
בָּרוּךְ מְרַחֵם עַל הַבְּרִיּוֹת,
בָּרוּךְ מְשַׁלֵּם שָׂכָר טוֹב לִירֵאָיו.
בָּרוּךְ חַי לָעַד וְקַיָּם לָנֶצַח.
בָּרוּךְ פּוֹדֶה וּמַצִּיל, בָּרוּךְ שְׁמוֹ.
בִּשְׁבָחוֹת וּבִזְמִרוֹת נְגַדֶּלְךָ
וּנְשַׁבֵּחֲךָ וּנְפָאֶרְךָ וְנַזְכִּיר
שִׁמְךָ וְנַמְלִיכְךָ, מַלְכֵּנוּ אֱלֹהֵינוּ.
יָחִיד, חֵי הָעוֹלָמִים, מֶלֶךְ מְשֻׁבָּח
וּמְפֹאָר, עֲדֵי עַד שְׁמוֹ הַגָּדוֹל.
בָּרוּךְ אַתָּה, יְיָ,
מֶלֶךְ מְהֻלָּל בַּתִּשְׁבָּחוֹת.

BLESSED is the One who spoke and the world came to be. Blessed is the One! Blessed is the One who continually authors creation. Blessed is the One whose word is deed; blessed is the One who decrees and fulfills. Blessed is the One who is compassionate towards the world; blessed is the One who is compassionate towards all creatures. Blessed is the One who rewards the reverent; blessed is the One who exists for all time, ever-enduring. Blessed is the One who redeems and saves; blessed is God's Name! With songs of praise, we extol You and proclaim Your Sovereignty, for You are the Source of life in the universe. One God, Life of the Universe, praised and glorious Ruler, Your Name is Eternal.

Blessed are You, Adonai, Sovereign who is glorified through praise.

בָּרוּךְ אַתָּה, יְיָ, מֶלֶךְ מְהֻלָּל בַּתִּשְׁבָּחוֹת.

פְּסוּקֵי דְזִמְרָה *P'sukei D'zimrah . . . Verses of Praise,* might be viewed as "prayer before prayer." It functions as the warm-up for the morning service, a recognition that prayerfulness cannot be summoned on demand. *Lawrence A. Hoffman*

Baruch She-amar

Psalm 92

Ashrei

Psalm 145

Psalms 150

Nishmat Kol Chai

Yishtabach

Chatzi Kaddish

בָּרוּךְ שֶׁאָמַר
מִזְמוֹר צ"ב
אַשְׁרֵי
מִזְמוֹר קמ"ה
מִזְמוֹר ק"נ
נִשְׁמַת כָּל חַי
יִשְׁתַּבַּח
חֲצִי קַדִּישׁ

This section through Yishtabach on page 223 offers no alternative readings.

PSALM 92:1–7, 13–16

מִזְמוֹר שִׁיר לְיוֹם הַשַּׁבָּת.
טוֹב לְהֹדוֹת לַייָ
וּלְזַמֵּר לְשִׁמְךָ עֶלְיוֹן.
לְהַגִּיד בַּבֹּקֶר חַסְדֶּךָ,
וֶאֱמוּנָתְךָ בַּלֵּילוֹת.
עֲלֵי־עָשׂוֹר וַעֲלֵי־נָבֶל,
עֲלֵי הִגָּיוֹן בְּכִנּוֹר.
כִּי שִׂמַּחְתַּנִי יְיָ בְּפָעֳלֶךָ
בְּמַעֲשֵׂי יָדֶיךָ אֲרַנֵּן.
מַה־גָּדְלוּ מַעֲשֶׂיךָ, יְיָ,
מְאֹד עָמְקוּ מַחְשְׁבֹתֶיךָ.
אִישׁ־בַּעַר לֹא יֵדָע
וּכְסִיל לֹא־יָבִין אֶת־זֹאת.
צַדִּיק כַּתָּמָר יִפְרָח
כְּאֶרֶז בַּלְּבָנוֹן יִשְׂגֶּה.
שְׁתוּלִים בְּבֵית יְיָ,
בְּחַצְרוֹת אֱלֹהֵינוּ יַפְרִיחוּ.
עוֹד יְנוּבוּן בְּשֵׂיבָה,
דְּשֵׁנִים וְרַעֲנַנִּים יִהְיוּ.
לְהַגִּיד כִּי־יָשָׁר יְיָ,
צוּרִי וְלֹא־עַוְלָתָה בּוֹ.

A PSALM, a song for Shabbat.
It is good to praise Adonai; to sing hymns to Your name, O Most High,
to proclaim Your steadfast love at daybreak, Your faithfulness each night
with a ten-stringed harp, with voice and lyre together.
You have gladdened me by Your deeds, Adonai; I shout for joy at Your handiwork.
How great are Your works, Adonai, how very subtle Your designs!
A brute cannot know, a fool cannot understand this:
The righteous bloom like a date-palm; they thrive like a cedar in Lebanon;
planted in the house of Adonai, they flourish in the courts of our God.
In old age they still produce fruit; they are full of sap and freshness,
attesting that Adonai is upright, my Rock, in whom there is no wrong.

אַשְׁרֵי יוֹשְׁבֵי בֵיתֶךָ,
עוֹד יְהַלְלְוּךָ סֶּלָה.
אַשְׁרֵי הָעָם שֶׁכָּכָה לּוֹ,
אַשְׁרֵי הָעָם שֶׁיְיָ אֱלֹהָיו.

HAPPY are those who dwell in Your house; they forever praise You!
Happy the people who have it so; happy the people whose God is Adonai.

DAVID'S PSALM — PSALM 145:1-21

תְּהִלָּה לְדָוִד.
אֲרוֹמִמְךָ אֱלוֹהַי הַמֶּלֶךְ,
וַאֲבָרְכָה שִׁמְךָ לְעוֹלָם וָעֶד.

בְּכָל־יוֹם אֲבָרְכֶךָּ,
וַאֲהַלְלָה שִׁמְךָ לְעוֹלָם וָעֶד.

גָּדוֹל יְיָ וּמְהֻלָּל מְאֹד,
וְלִגְדֻלָּתוֹ אֵין חֵקֶר.

דּוֹר לְדוֹר יְשַׁבַּח מַעֲשֶׂיךָ,
וּגְבוּרֹתֶיךָ יַגִּידוּ.

הֲדַר כְּבוֹד הוֹדֶךָ,
וְדִבְרֵי נִפְלְאֹתֶיךָ אָשִׂיחָה.

וֶעֱזוּז נוֹרְאֹתֶיךָ יֹאמֵרוּ,
וּגְדֻלָּתְךָ אֲסַפְּרֶנָּה.

זֵכֶר רַב־טוּבְךָ יַבִּיעוּ,
וְצִדְקָתְךָ יְרַנֵּנוּ.

חַנּוּן וְרַחוּם יְיָ,
אֶרֶךְ אַפַּיִם וּגְדָל־חָסֶד.

טוֹב יְיָ לַכֹּל,
וְרַחֲמָיו עַל־כָּל־מַעֲשָׂיו.

יוֹדוּךָ יְיָ כָּל־מַעֲשֶׂיךָ,
וַחֲסִידֶיךָ יְבָרְכוּכָה.

אַשְׁרֵי יוֹשְׁבֵי *Ashrei yoshvei . . . Happy are those who dwell . . .* Psalm 84:5
אַשְׁרֵי הָעָם *Ashrei haam . . . Happy the people . . .* Psalm 144:15

בָּרוּךְ שֶׁאָמַר
מִזְמוֹר צ"ב
אַשְׁרֵי
מִזְמוֹר קמ"ה
מִזְמוֹר ק"נ
נִשְׁמַת כָּל חַי
יִשְׁתַּבַּח
חֲצִי קַדִּישׁ

כְּבוֹד מַלְכוּתְךָ יֹאמֵרוּ,
וּגְבוּרָתְךָ יְדַבֵּרוּ.

לְהוֹדִיעַ לִבְנֵי הָאָדָם גְּבוּרֹתָיו,
וּכְבוֹד הֲדַר מַלְכוּתוֹ.

מַלְכוּתְךָ מַלְכוּת כָּל־עֹלָמִים,
וּמֶמְשַׁלְתְּךָ בְּכָל־דּוֹר וָדוֹר.

סוֹמֵךְ יְיָ לְכָל־הַנֹּפְלִים,
וְזוֹקֵף לְכָל־הַכְּפוּפִים.

עֵינֵי כֹל אֵלֶיךָ יְשַׂבֵּרוּ,
וְאַתָּה נוֹתֵן־לָהֶם אֶת־אָכְלָם בְּעִתּוֹ.

פּוֹתֵחַ אֶת־יָדֶךָ,
וּמַשְׂבִּיעַ לְכָל־חַי רָצוֹן.

צַדִּיק יְיָ בְּכָל־דְּרָכָיו,
וְחָסִיד בְּכָל־מַעֲשָׂיו.

קָרוֹב יְיָ לְכָל־קֹרְאָיו,
לְכֹל אֲשֶׁר יִקְרָאֻהוּ בֶאֱמֶת.

רְצוֹן־יְרֵאָיו יַעֲשֶׂה,
וְאֶת־שַׁוְעָתָם יִשְׁמַע וְיוֹשִׁיעֵם.

שׁוֹמֵר יְיָ אֶת־כָּל־אֹהֲבָיו,
וְאֵת כָּל־הָרְשָׁעִים יַשְׁמִיד.

תְּהִלַּת יְיָ יְדַבֶּר־פִּי,
וִיבָרֵךְ כָּל־בָּשָׂר שֵׁם קָדְשׁוֹ
לְעוֹלָם וָעֶד.

וַאֲנַחְנוּ נְבָרֵךְ יָהּ
מֵעַתָּה וְעַד עוֹלָם, הַלְלוּיָהּ.

וַאֲנַחְנוּ נְבָרֵךְ יָהּ *Vaanachnu n'vareich Yah . . . and all creatures shall bless . . .* Psalm 115:18

DAVID'S SONG OF PRAISE

I will extol You, my God and Sovereign,
and bless Your name forever and ever.
Every day will I bless You and praise Your name forever and ever.
Great is Adonai and much acclaimed;
God's greatness cannot be fathomed.
One generation shall laud Your works to another
and declare Your mighty acts.
The glorious majesty of Your splendor
and Your wondrous acts will I recite.
They shall talk of the might of Your awesome deeds,
and I will recount Your greatness.
They shall celebrate Your abundant goodness,
and sing joyously of Your beneficence.
Adonai is gracious and compassionate,
slow to anger and abounding in kindness.
Adonai is good to all, and God's mercy is upon all God's works.
All Your works shall praise You, Adonai,
and Your faithful ones shall bless You.
They shall talk of the majesty of Your sovereignty,
and speak of Your might,
to make God's mighty acts known among mortals
and the majestic glory of Your sovereignty.
Your sovereignty is eternal; Your dominion is for all generations.
Adonai supports all who stumble,
and makes all who are bent stand straight.
The eyes of all look to You expectantly,
and You give them their food when it is due.
You give it openhandedly,
feeding every creature to its heart's content.
Adonai is beneficent in all ways and faithful in all works.
Adonai is near to all who call
to all who call upon God with sincerity.
Adonai fulfills the wishes of those who fear God;
Adonai hears their cry and delivers them.
Adonai watches over all who love God,
but all the wicked God will destroy.
My mouth shall utter the praise of Adonai,
and all creatures shall bless God's holy name forever and ever.
We will bless You God now and always. Hallelujah!

בָּרוּךְ שֶׁאָמַר

מִזְמוֹר צ"ב

אַשְׁרֵי

מִזְמוֹר קמ"ה

מִזְמוֹר ק"נ

נִשְׁמַת כָּל חַי

יִשְׁתַּבַּח

חֲצִי קַדִּישׁ

Baruch She-amar
Psalm 92
Ashrei
Psalm 145
Psalms 150
Nishmat Kol Chai
Yishtabach
Chatzi Kaddish

PSALM 150:1–6

הַלְלוּ יָהּ!

הַלְלוּ־אֵל בְּקָדְשׁוֹ,
הַלְלוּהוּ בִּרְקִיעַ עֻזּוֹ.
הַלְלוּהוּ בִגְבוּרֹתָיו,
הַלְלוּהוּ כְּרֹב גֻּדְלוֹ.
הַלְלוּהוּ בְּתֵקַע שׁוֹפָר,
הַלְלוּהוּ בְּנֵבֶל וְכִנּוֹר.
הַלְלוּהוּ בְּתֹף וּמָחוֹל,
הַלְלוּהוּ בְּמִנִּים וְעוּגָב.
הַלְלוּהוּ בְצִלְצְלֵי־שָׁמַע,
הַלְלוּהוּ בְּצִלְצְלֵי־תְרוּעָה.
כֹּל הַנְּשָׁמָה תְּהַלֵּל יָהּ,
הַלְלוּ־יָהּ!

Hallelujah!

Praise God in God's sanctuary;
praise God in the sky, God's stronghold.
Praise God for mighty acts;
praise God for God's exceeding greatness.
Praise God with blasts of the horn;
praise God with harp and lyre.
Praise God with timbrel and dance;
praise God with lute and pipe.
Praise God with resounding cymbals;
praise God with loud-clashing cymbals.
Let all that breathes praise God.
Hallelujah!

נִשְׁמַת כָּל חַי
תְּבָרֵךְ אֶת שִׁמְךָ, יְיָ אֱלֹהֵינוּ,
וְרוּחַ כָּל בָּשָׂר
תְּפָאֵר וּתְרוֹמֵם זִכְרְךָ,
מַלְכֵּנוּ, תָּמִיד.
מִן הָעוֹלָם וְעַד הָעוֹלָם אַתָּה אֵל,
אֵין לָנוּ מֶלֶךְ אֶלָּא אָתָּה.
אֱלֹהֵי הָרִאשׁוֹנִים וְהָאַחֲרוֹנִים,
אֱלוֹהַּ כָּל בְּרִיּוֹת, אֲדוֹן כָּל תּוֹלָדוֹת,
הַמְהֻלָּל בְּרֹב הַתִּשְׁבָּחוֹת,
הַמְנַהֵג עוֹלָמוֹ בְּחֶסֶד
וּבְרִיּוֹתָיו בְּרַחֲמִים.
וַייָ לֹא יָנוּם וְלֹא יִישָׁן.
הַמְעוֹרֵר יְשֵׁנִים וְהַמֵּקִיץ נִרְדָּמִים,
וְהַמֵּשִׂיחַ אִלְּמִים, וְהַמַּתִּיר אֲסוּרִים,
וְהַסּוֹמֵךְ נוֹפְלִים, וְהַזּוֹקֵף כְּפוּפִים.
לְךָ לְבַדְּךָ אֲנַחְנוּ מוֹדִים.

LET THE SOUL of everything alive bless Your name, Adonai, our God;
and the spirit of all flesh glorify and exalt Your name forever, O Sovereign.
Transcending space and time, You alone are God.
We have no Sovereign besides You.
God of the first and the last, God of all creatures,
Master of all generations, who is praised in a multitude of praises,
who guides the world with abundant loving-kindness,
and all creatures with mercy. Adonai neither slumbers nor sleeps.
God awakens the sleeping, arouses those who slumber,
gives speech to the mute;
and God loosens the bonds of captives, God supports the fallen,
and strengthens those who are bent over.
You alone do we acknowledge.

מִן הָעוֹלָם וְעַד הָעוֹלָם *Min haolam v'ad haolam . . . Transcending space and time . . . based on* Psalm 90:2
וַייָ לֹא יָנוּם *VaAdonai lo yanum . . . Adonai neither slumbers . . . based on* Psalm 121:4
וְהַמַּתִּיר אֲסוּרִים *v'hamatir asurim . . . loosens the bonds of captives . . . based on* Psalm 146:7
וְהַסּוֹמֵךְ נוֹפְלִים *v'hasomeich noflim . . . supports the fallen . . . based on* Psalm 145:14

בָּרוּךְ שֶׁאָמַר

מִזְמוֹר צ"ב

אַשְׁרֵי

מִזְמוֹר קמ"ה

מִזְמוֹר ק"נ

נִשְׁמַת כָּל חַי

יִשְׁתַּבַּח

חֲצִי קַדִּישׁ

אִלּוּ פִינוּ מָלֵא שִׁירָה כַּיָּם,

וּלְשׁוֹנֵנוּ רִנָּה כַּהֲמוֹן גַּלָּיו,

וְשִׂפְתוֹתֵינוּ שֶׁבַח כְּמֶרְחֲבֵי רָקִיעַ,

וְעֵינֵינוּ מְאִירוֹת כַּשֶּׁמֶשׁ וְכַיָּרֵחַ,

וְיָדֵינוּ פְרוּשׂוֹת כְּנִשְׁרֵי שָׁמָיִם,

וְרַגְלֵינוּ קַלּוֹת כָּאַיָּלוֹת,

אֵין אֲנַחְנוּ מַסְפִּיקִים לְהוֹדוֹת לָךְ,

יְיָ אֱלֹהֵינוּ וֵאלֹהֵי

אֲבוֹתֵינוּ וְאִמּוֹתֵינוּ,

וּלְבָרֵךְ אֶת שְׁמֶךָ

עַל אַחַת מֵאֶלֶף, אֶלֶף אַלְפֵי אֲלָפִים

וְרִבֵּי רְבָבוֹת פְּעָמִים

הַטּוֹבוֹת שֶׁעָשִׂיתָ

עִם אֲבוֹתֵינוּ וְאִמּוֹתֵינוּ וְעִמָּנוּ.

EVEN IF OUR MOUTHS were full of song as the sea,
  and our tongues full of joy in countless waves,
and our lips full of praise as wide as the sky's expanse,
  and were our eyes to shine like sun and moon;
if our hands were spread out like heaven's eagles
  and our feet swift like young deer,
we could never thank You adequately, Adonai,
  our God and God of our ancestors,
    to bless Your Name for a ten-thousandth
    of the many myriads of times
You granted favors to our ancestors and to us.

Every praise could be music, the voice singing in harmony with the universe and its Creator!

עַל כֵּן אֵבָרִים שֶׁפִּלַּגְתָּ בָּנוּ,
וְרוּחַ וּנְשָׁמָה שֶׁנָּפַחְתָּ בְּאַפֵּינוּ,
וְלָשׁוֹן אֲשֶׁר שַׂמְתָּ בְּפִינוּ,
הֵן הֵם יוֹדוּ וִיבָרְכוּ וִישַׁבְּחוּ
וִיפָאֲרוּ אֶת שִׁמְךָ, מַלְכֵּנוּ.
כִּי כָל פֶּה לְךָ יוֹדֶה,
וְכָל לָשׁוֹן לְךָ תִשָּׁבַע,
וְכָל בֶּרֶךְ לְךָ תִכְרַע,
וְכָל קוֹמָה לְפָנֶיךָ תִשְׁתַּחֲוֶה,
וְכָל לְבָבוֹת יִירָאוּךָ,
וְכָל קֶרֶב וּכְלָיוֹת יְזַמְּרוּ לִשְׁמֶךָ,
כַּדָּבָר שֶׁכָּתוּב: כָּל עַצְמוֹתַי תֹּאמַרְנָה:
יְיָ, מִי כָמוֹךָ.
כָּאָמוּר: לְדָוִד, בָּרְכִי נַפְשִׁי אֶת יְיָ,
וְכָל קְרָבַי אֶת שֵׁם קָדְשׁוֹ.

THEREFORE THESE LIMBS which You have formed in us,
and this spirit and soul that You breathed into our nostrils,
this tongue which You have set in our mouths,
they must acknowledge, bless, praise and glorify
Your Name, O our Sovereign.
For every mouth will acknowledge You
and every tongue pledge homage to You,
every knee bend in Your presence,
every upright person shall bow before You.
Every heart will revere You
and every inmost thought will sing to Your Name.
And David said:
"Let all my bones exclaim: 'Who is like You, Adonai?' "
As David continued:
"Bless Adonai, O my soul, and let my inner being praise Your holy Name."

וְכָל בֶּרֶךְ *v'chol berech . . . every knee . . . based on* Isaiah 45:23

כָּל עַצְמוֹתַי תֹּאמַרְנָה *kol atzmotai tomarnah . . . Let all my bones exclaim . . .* Psalm 35:10

בָּרְכִי נַפְשִׁי אֶת יְיָ *bar'chi nafshi et Adonai . . . Bless Adonai, O my soul . . .* Psalm 103:1

בָּרוּךְ שֶׁאָמַר

מִזְמוֹר צ"ב

אַשְׁרֵי

מִזְמוֹר קמ"ה

מִזְמוֹר ק"ן

נִשְׁמַת כָּל חַי

יִשְׁתַּבַּח

חֲצִי קַדִּישׁ

הָאֵל בְּתַעֲצֻמוֹת עֻזֶּךָ,

הַגָּדוֹל בִּכְבוֹד שְׁמֶךָ,

הַגִּבּוֹר לָנֶצַח

וְהַנּוֹרָא בְּנוֹרְאוֹתֶיךָ,

הַמֶּלֶךְ הַיּוֹשֵׁב

עַל כִּסֵּא רָם וְנִשָּׂא.

שׁוֹכֵן עַד, מָרוֹם וְקָדוֹשׁ שְׁמוֹ.

וְכָתוּב: רַנְּנוּ צַדִּיקִים בַּיָי,

לַיְשָׁרִים נָאוָה תְהִלָּה.

בְּפִי יְשָׁרִים תִּתְהַלָּל.

וּבְדִבְרֵי צַדִּיקִים תִּתְבָּרַךְ.

וּבִלְשׁוֹן חֲסִידִים תִּתְרוֹמָם.

וּבְקֶרֶב קְדוֹשִׁים תִּתְקַדָּשׁ.

וּבְמַקְהֲלוֹת רִבְבוֹת עַמְּךָ, בֵּית יִשְׂרָאֵל,

בְּרִנָּה יִתְפָּאַר שִׁמְךָ מַלְכֵּנוּ

בְּכָל דּוֹר וָדוֹר.

O GOD, IN THE POWER of Your strength,
great in the glory of Your Name,
Mighty forever, Awesome in amazing deeds,
the Ruler who sits on a high and exalted throne.

O God, Your name is holy and exalted.
    The Psalmist has said:
"Rejoice in Adonai, you righteous;
    praise suits the upright."
In the mouths of the upright You will be praised,
    and in the words of the righteous You will be blessed.
In the tongue of the faithful You will be exalted
    and in the midst of the holy You will be sanctified.

In the multitude of choirs of Your people, the House of Israel,
    Your name is to be glorified in joy in every generation, O our Sovereign.

הָאֵל בְּתַעֲצֻמוֹת עֻזֶּךָ *Ha-El b'taatzumot uzecha . . . O God, in the power . . . based on* Psalm 68:36

רָם וְנִשָּׂא. שׁוֹכֵן *ram v'nisa. Shochein . . . based on* Isaiah 57:15

רַנְּנוּ צַדִּיקִים בַּיָי *Ran'nu tzadikim b'Adonai . . . Rejoice in Adonai, you righteous . . .* Psalm 33:1

יִשְׁתַּבַּח שִׁמְךָ לָעַד מַלְכֵּנוּ,
הָאֵל הַמֶּלֶךְ הַגָּדוֹל וְהַקָּדוֹשׁ
בַּשָּׁמַיִם וּבָאָרֶץ.
כִּי לְךָ נָאֶה, יְיָ אֱלֹהֵינוּ
וֵאלֹהֵי אֲבוֹתֵינוּ וְאִמּוֹתֵינוּ,
שִׁיר וּשְׁבָחָה, הַלֵּל וְזִמְרָה,
עֹז וּמֶמְשָׁלָה, נֶצַח
גְּדֻלָּה וּגְבוּרָה,
תְּהִלָּה וְתִפְאֶרֶת, קְדֻשָּׁה וּמַלְכוּת,
בְּרָכוֹת וְהוֹדָאוֹת מֵעַתָּה וְעַד עוֹלָם.

YOU SHALL ALWAYS BE PRAISED,
great and holy God, our Sovereign in heaven and on earth.
Songs of praise and psalms of adoration become You,
acknowledging Your might and Your dominion.
Yours are strength and sovereignty, sanctity, grandeur and glory always.
We offer our devotion, open our hearts in acclamation.

בָּרוּךְ אַתָּה, יְיָ,
אֵל מֶלֶךְ גָּדוֹל בַּתִּשְׁבָּחוֹת,
אֵל הַהוֹדָאוֹת, אֲדוֹן הַנִּפְלָאוֹת,
הַבּוֹחֵר בְּשִׁירֵי זִמְרָה,
מֶלֶךְ אֵל חֵי הָעוֹלָמִים.

Praised are You, Sovereign of wonders,
crowned in adoration,
delighting in song,
Eternal Majesty.

The first half of יִשְׁתַּבַּח *Yishtabach* contains 15 expressions of praise for God. The last sentence (after בָּרוּךְ אַתָּה, יְיָ *Baruch atah, Adonai*) contains 15 words. Some consider this an allusion to one of the Divine Names, יה, whose numerical value is 15. When we speak the 15 words, we metaphorically praise God's Name.

בָּרוּךְ שֶׁאָמַר

מִזְמוֹר צ״ב

אַשְׁרֵי

מִזְמוֹר קמ״ה

מִזְמוֹר ק״נ

נִשְׁמַת כָּל חַי

יִשְׁתַּבַּח

חֲצִי קַדִּישׁ

בָּרוּךְ שֶׁאָמַר
מִזְמוֹר צ״ב
אַשְׁרֵי
מִזְמוֹר קמ״ה
מִזְמוֹר ק״נ
נִשְׁמַת כָּל חַי
יִשְׁתַּבַּח
חֲצִי קַדִּישׁ

יִתְגַּדַּל וְיִתְקַדַּשׁ שְׁמֵהּ רַבָּא
בְּעָלְמָא דִּי בְרָא כִרְעוּתֵהּ,
וְיַמְלִיךְ מַלְכוּתֵהּ
בְּחַיֵּיכוֹן וּבְיוֹמֵיכוֹן
וּבְחַיֵּי דְכָל בֵּית יִשְׂרָאֵל,
בַּעֲגָלָא וּבִזְמַן קָרִיב,
וְאִמְרוּ: אָמֵן.

יְהֵא שְׁמֵהּ רַבָּא מְבָרַךְ
לְעָלַם וּלְעָלְמֵי עָלְמַיָּא.

יִתְבָּרַךְ וְיִשְׁתַּבַּח וְיִתְפָּאַר
וְיִתְרוֹמַם וְיִתְנַשֵּׂא,
וְיִתְהַדַּר וְיִתְעַלֶּה וְיִתְהַלַּל
שְׁמֵהּ דְּקֻדְשָׁא בְּרִיךְ הוּא,
לְעֵלָּא מִן כָּל בִּרְכָתָא וְשִׁירָתָא,
תֻּשְׁבְּחָתָא וְנֶחֱמָתָא,
דַּאֲמִירָן בְּעָלְמָא, וְאִמְרוּ: אָמֵן.

EXALTED and hallowed be God's great name,
in the world which God created, according to plan.
May God's majesty be revealed in the days of our lifetime
and the life of all Israel —
speedily, imminently.
To which we say: Amen.

Blessed be God's great name to all eternity.

Blessed, praised, honored, exalted,
extolled, glorified, adored, and lauded
be the name of the Holy Blessed One,
beyond all earthly words and songs of blessing, praise, and comfort.
To which we say: Amen.

God can hardly be listening to the actual words of our prayers: how unendingly boring to hear such repetition days without number! And anyone who prays with any regularity or sincerity knows that the same prayers mean different things on different readings. Instead, each prayer is a metaphoric representation of the speaker's heart and mind. Now to God, that is an interesting and unending text, ever-changing, and God is the ultimate reader of nuance and allusion. *Adam Sol*

E The *Chatzi Kaddish* is in the format of a call and response. The *sh'liach tzibur* (prayer leader) and the congregation trade parts, with the *sh'liach tzibur* using the formula "To which we say" (*v'imru*) in order to invite the congregation into the prayer, with the congregation responding, "Amen." How is a good worship service the product of the interaction between the leaders and the congregation? What is the synergy between the two that creates the "magic" of a prayer service?

בָּרְכוּ

יוֹצֵר

אַהֲבָה רַבָּה

שְׁמַע

וְאָהַבְתָּ

לְמַעַן תִּזְכְּרוּ

וַיֹּאמֶר יְיָ

אֱמֶת וְיַצִּיב

מִי־כָמֹכָה

שְׁמַע וּבִרְכוֹתֶיהָ

SH'MA UVIRCHOTEHA — SH'MA AND ITS BLESSINGS

בָּרְכוּ אֶת יְיָ הַמְבֹרָךְ!
בָּרוּךְ יְיָ הַמְבֹרָךְ
לְעוֹלָם וָעֶד!

PRAISE ADONAI to whom praise is due forever!
Praised be Adonai to whom praise is due,
now and forever!

For those who choose: The prayer leader at the word בָּרְכוּ *Bar'chu* (the call to worship) bends the knees and bows from the waist, and at יְיָ *Adonai* stands straight. בָּרוּךְ יְיָ *Baruch Adonai* is the communal response, whereupon the community repeats the choreography of the first line.

Bar'chu

Yotzeir

Ahavah Rabbah

Sh'ma

V'ahavta

L'maan tizk'ru

Vayomer Adonai

Emet v'Yatziv

Mi Chamochah

בָּרוּךְ אַתָּה, יְיָ
אֱלֹהֵינוּ, מֶלֶךְ הָעוֹלָם,
יוֹצֵר אוֹר וּבוֹרֵא חְשֶׁךְ,
עֹשֶׂה שָׁלוֹם וּבוֹרֵא אֶת־הַכֹּל.
הַמֵּאִיר לָאָרֶץ
וְלַדָּרִים עָלֶיהָ בְּרַחֲמִים,
וּבְטוּבוֹ מְחַדֵּשׁ בְּכָל יוֹם תָּמִיד
מַעֲשֵׂה בְרֵאשִׁית.
מָה רַבּוּ מַעֲשֶׂיךָ, יְיָ,
כֻּלָּם בְּחָכְמָה עָשִׂיתָ,
מָלְאָה הָאָרֶץ קִנְיָנֶךָ.
תִּתְבָּרַךְ, יְיָ אֱלֹהֵינוּ,
עַל שֶׁבַח מַעֲשֵׂה יָדֶיךָ
וְעַל מְאוֹרֵי אוֹר שֶׁעָשִׂיתָ,
יְפָאֲרוּךָ סֶּלָה.
אוֹר חָדָשׁ עַל צִיּוֹן תָּאִיר,
וְנִזְכֶּה כֻלָּנוּ מְהֵרָה לְאוֹרוֹ.
בָּרוּךְ אַתָּה, יְיָ, יוֹצֵר הַמְּאוֹרוֹת.

בָּרְכוּ

יוֹצֵר

אַהֲבָה רַבָּה

שְׁמַע

וְאָהַבְתָּ

לְמַעַן תִּזְכְּרוּ

וַיֹּאמֶר יְיָ

אֱמֶת וְיַצִּיב

מִי־כָמֹכָה

PRAISED ARE YOU, Adonai our God, Sovereign of the universe,
Creator of light and darkness, who makes peace and fashions all things.
In mercy, You illumine the world and those who live upon it.
In Your goodness You daily renew creation.
How numerous are Your works, Adonai!
In wisdom, You formed them all, filling the earth with Your creatures.
Be praised, Adonai our God, for the excellent work of Your hands,
and for the lights You created, may they glorify You.
Shine a new light upon Zion, that we all may swiftly merit its radiance.
Praised are You, Adonai, Creator of all heavenly lights.

בָּרוּךְ אַתָּה, יְיָ, יוֹצֵר הַמְּאוֹרוֹת.

אוֹר חָדָשׁ עַל צִיּוֹן תָּאִיר *Or chadash al Tzion ta-ir . . .   Shine a new light upon Zion . . .*
Classical Reform prayerbook authors in the Diaspora consistently omitted this line with its
mention of Zion from the liturgy because of their opposition to Jewish nationalism. With the
restoration of this passage to *Mishkan T'filah,* our movement consciously affirms its devotion to
the modern State of Israel and signals its recognition of the religious significance of the reborn
Jewish commonwealth. *David Ellenson*

בָּרוּךְ אַתָּה, יְיָ . . . יוֹצֵר אוֹר וּבוֹרֵא חְשֶׁךְ *Baruch atah, Adonai . . . yotzeir or uvorei choshech . . .*
Praised are You, Adonai . . . Who forms light and creates darkness . . .   based on Isaiah 45:7

מָה רַבּוּ *Mah rabu . . .   How numerous . . .* Psalm 104:24

P Throughout the history of the Reform Movement, changes have been made in the prayers in order to remove beliefs that we no longer feel are meaningful. In the last two prayerbooks, *The Union Prayer Book* and *Gates of Prayer,* the line אוֹר חָדָשׁ עַל צִיּוֹן תָּאִיר *Or chadash al Tzion ta-ir* ("Shine a new light upon Zion [Israel]") was taken out. However, the editors of *Mishkan T'filah* chose to put this line back into the *Yotzeir.* Why do you think they made this choice? What new light do you think they are envisioning?

בָּרְכוּ

יוֹצֵר

אַהֲבָה רַבָּה

שְׁמַע

וְאָהַבְתָּ

לְמַעַן תִּזְכְּרוּ

וַיֹּאמֶר יְיָ

אֱמֶת וְיַצִּיב

מִי־כָמְכָה

אַהֲבָה רַבָּה אֲהַבְתָּנוּ, יְיָ אֱלֹהֵינוּ,
חֶמְלָה גְדוֹלָה וִיתֵרָה חָמַלְתָּ עָלֵינוּ.
בַּעֲבוּר אֲבוֹתֵינוּ וְאִמּוֹתֵינוּ שֶׁבָּטְחוּ בְךָ
וַתְּלַמְּדֵם חֻקֵּי חַיִּים, כֵּן תְּחָנֵּנוּ
וּתְלַמְּדֵנוּ. הַמְרַחֵם, רַחֵם עָלֵינוּ,
וְתֵן בְּלִבֵּנוּ לְהָבִין וּלְהַשְׂכִּיל, לִשְׁמֹעַ,
לִלְמֹד וּלְלַמֵּד, לִשְׁמֹר וְלַעֲשׂוֹת וּלְקַיֵּם
אֶת־כָּל־דִּבְרֵי תַלְמוּד תּוֹרָתֶךָ בְּאַהֲבָה.

HOW DEEPLY You have loved us Adonai, our God, gracing us with surpassing compassion! On account of our forebears whose trust led You to teach them the laws of life, be gracious to us, teaching us as well. O Merciful One, have mercy on us by making us able to understand and discern, to heed, learn, and teach, and, lovingly, to observe, perform, and fulfill all that is in Your Torah.

וְהָאֵר עֵינֵינוּ בְּתוֹרָתֶךָ,
וְדַבֵּק לִבֵּנוּ בְּמִצְוֹתֶיךָ,
וְיַחֵד לְבָבֵנוּ לְאַהֲבָה
וּלְיִרְאָה אֶת־שְׁמֶךָ,
וְלֹא נֵבוֹשׁ וְלֹא נִכָּלֵם,
וְלֹא נִכָּשֵׁל לְעוֹלָם וָעֶד.
כִּי בְשֵׁם קָדְשְׁךָ הַגָּדוֹל וְהַנּוֹרָא
בָּטָחְנוּ, נָגִילָה וְנִשְׂמְחָה בִּישׁוּעָתֶךָ.
וַהֲבִיאֵנוּ לְשָׁלוֹם מֵאַרְבַּע כַּנְפוֹת
הָאָרֶץ, וְתוֹלִיכֵנוּ קוֹמְמִיּוּת לְאַרְצֵנוּ.
כִּי אֵל פּוֹעֵל יְשׁוּעוֹת אָתָּה, וּבָנוּ בָחַרְתָּ
וְקֵרַבְתָּנוּ לְשִׁמְךָ הַגָּדוֹל סֶלָה בֶּאֱמֶת,
לְהוֹדוֹת לְךָ וּלְיַחֶדְךָ בְּאַהֲבָה.
בָּרוּךְ אַתָּה, יְיָ,
הַבּוֹחֵר בְּעַמּוֹ יִשְׂרָאֵל בְּאַהֲבָה.

Enlighten our eyes with Your Torah, focus our minds on Your mitzvot, unite our hearts in love and reverence for Your Name. Then we will never feel shame, never deserve rebuke, and never stumble. Having trusted in Your great and awesome holiness, we shall celebrate Your salvation with joy.

Gather us in peace from the four corners of the earth and lead us upright to our land. For You, O God, work wonders. You chose us. Truly, You drew us near to Your Great Name, that we might acknowledge You, declaring You One in love. Praised be You, Adonai, who chooses Your people Israel in love.

בָּרוּךְ אַתָּה, יְיָ, הַבּוֹחֵר בְּעַמּוֹ יִשְׂרָאֵל בְּאַהֲבָה.
Baruch atah, Adonai, habocheir b'amo Yisrael b'ahavah.

Bar'chu
Yotzeir
Ahavah Rabbah
Sh'ma
V'ahavta
L'maan tizk'ru
Vayomer Adonai
Emet v'Yatziv
Mi Chamochah

P Parents say that they provide rules and restrictions out of love. God's great love is shown in this prayer by Revelation — the giving of the covenant at Sinai, the rules of the Ten Commandments. Why do you think setting down rules is a way to show love? How is God's love similar to or different from parental love?

For those who choose: At the words וַהֲבִיאֵנוּ לְשָׁלוֹם *V'havi-einu l'shalom, Gather us in peace,* one gathers the four fringes of the tallit in the left hand and holds them throughout the שְׁמַע *Sh'ma* to symbolize the ingathering of our people.

שמע ישראל יהוה

שְׁמַ֖ע יִשְׂרָאֵ֑ל יְהוָ֥ה אֱלֹהֵ֖ינוּ יְהוָ֥ה אֶחָֽד!

Hear, O Israel, Adonai is our God, Adonai is One!

בָּרוּךְ שֵׁם כְּבוֹד מַלְכוּתוֹ לְעוֹלָם וָעֶד.

Blessed is God's glorious majesty forever and ever.

בָּרְכוּ
יוֹצֵר
אַהֲבָה רַבָּה
שְׁמַע
וְאָהַבְתָּ
לְמַעַן תִּזְכְּרוּ
וַיֹּאמֶר יְיָ
אֱמֶת וְיַצִּיב
מִי־כָמֹכָה

וְאָהַבְתָּ אֵת יְיָ אֱלֹהֶיךָ
בְּכָל־לְבָבְךָ וּבְכָל־נַפְשְׁךָ וּבְכָל־
מְאֹדֶךָ: וְהָיוּ הַדְּבָרִים הָאֵלֶּה
אֲשֶׁר אָנֹכִי מְצַוְּךָ הַיּוֹם עַל־
לְבָבֶךָ: וְשִׁנַּנְתָּם לְבָנֶיךָ וְדִבַּרְתָּ
בָּם בְּשִׁבְתְּךָ בְּבֵיתֶךָ וּבְלֶכְתְּךָ
בַדֶּרֶךְ וּבְשָׁכְבְּךָ וּבְקוּמֶךָ:
וּקְשַׁרְתָּם לְאוֹת עַל־יָדֶךָ וְהָיוּ
לְטֹטָפֹת בֵּין עֵינֶיךָ: וּכְתַבְתָּם
עַל־מְזֻזוֹת בֵּיתֶךָ וּבִשְׁעָרֶיךָ:

YOU SHALL LOVE Adonai your God with all your heart,
with all your soul, and with all your might.
Take to heart these instructions with which I charge you this day.
Impress them upon your children.
Recite them when you stay at home and when you are away,
when you lie down and when you get up.
Bind them as a sign on your hand and let them serve as a symbol on your forehead;
inscribe them on the doorposts of your house and on your gates.

Continue or turn to pages 236–237.

לְמַעַן תִּזְכְּרוּ וַעֲשִׂיתֶם אֶת־
כָּל־מִצְוֺתָי וִהְיִיתֶם קְדֹשִׁים
לֵאלֹהֵיכֶם: אֲנִי יְיָ אֱלֹהֵיכֶם אֲשֶׁר
הוֹצֵאתִי אֶתְכֶם מֵאֶרֶץ
מִצְרַיִם לִהְיוֹת לָכֶם לֵאלֹהִים
אֲנִי יְיָ אֱלֹהֵיכֶם:

Thus you shall remember to observe all My commandments
and to be holy to your God.
I am Adonai, your God, who brought you out of the land of Egypt to be your God:
I am Adonai your God.

יְיָ אֱלֹהֵיכֶם אֱמֶת.

Turn to pages 238–239.

For those who choose: At the end of the שְׁמַע *Sh'ma*, after the words יְיָ אֱלֹהֵיכֶם *Adonai Eloheichem,* the word אֱמֶת *emet* ("true") is added as an immediate affirmation of its truth.

וְאָהַבְתָּ *V'ahavta . . . You shall love . . .* Deuteronomy 6:5–9

לְמַעַן תִּזְכְּרוּ *L'maan tizk'ru . . . Thus you shall remember . . .* Numbers 15:40–41

Bar'chu
Yotzeir
Ahavah Rabbah
Sh'ma
V'ahavta
L'maan tizk'ru
Vayomer Adonai
Emet v'Yatziv
Mi Chamochah

P It seems counterintuitive to command love, yet that is the purpose of this prayer — to take something that we think of as a spontaneous emotion and require it for a presence that we cannot touch or see. Would we forget to love God without this command?

וַיֹּ֣אמֶר יְיָ אֶל־מֹשֶׁה לֵּאמֹֽר:
דַּבֵּ֞ר אֶל־בְּנֵ֤י יִשְׂרָאֵל֙
וְאָמַרְתָּ֣ אֲלֵהֶ֔ם וְעָשׂ֨וּ לָהֶ֥ם צִיצִ֛ת
עַל־כַּנְפֵ֥י בִגְדֵיהֶ֖ם לְדֹרֹתָ֑ם
וְנָתְנ֛וּ עַל־צִיצִ֥ת הַכָּנָ֖ף פְּתִ֥יל תְּכֵֽלֶת:
וְהָיָ֣ה לָכֶם֮ לְצִיצִת֒ וּרְאִיתֶ֣ם אֹת֗וֹ
וּזְכַרְתֶּם֙ אֶת־כָּל־מִצְוֺ֣ת יְיָ֔
וַעֲשִׂיתֶ֖ם אֹתָ֑ם
וְלֹֽא־תָת֜וּרוּ אַחֲרֵ֤י לְבַבְכֶם֙
וְאַחֲרֵ֣י עֵֽינֵיכֶ֔ם
אֲשֶׁר־אַתֶּ֥ם זֹנִ֖ים אַחֲרֵיהֶֽם:

בָּרְכוּ
יוֹצֵר
אַהֲבָה רַבָּה
שְׁמַע
וְאָהַבְתָּ
לְמַעַן תִּזְכְּרוּ
וַיֹּאמֶר יְיָ
אֱמֶת וְיַצִּיב
מִי־כָמֹֽכָה

ADONAI SAID TO MOSES as follows:
Speak to the Israelite people and instruct them to make for themselves
fringes on the corners of their garments throughout the ages;
let them attach a cord of blue to the fringe at each corner.
That shall be your fringe; look at it and recall all the commandments of Adonai
and observe them, so that you do not follow your heart and eyes in your lustful urge.

לְמַעַן תִּזְכְּר֔וּ וַעֲשִׂיתֶ֖ם אֶת־
כָּל־מִצְוֺתָ֑י וִהְיִיתֶ֥ם קְדֹשִׁ֖ים
לֵאלֹֽהֵיכֶֽם: אֲנִ֞י יְיָ֣ אֱלֹֽהֵיכֶ֗ם אֲשֶׁ֨ר
הוֹצֵ֤אתִי אֶתְכֶם֙ מֵאֶ֣רֶץ
מִצְרַ֔יִם לִהְי֥וֹת לָכֶ֖ם לֵֽאלֹהִ֑ים
אֲנִ֖י יְיָ֥ אֱלֹֽהֵיכֶֽם:

Thus you shall remember to observe all My commandments
and to be holy to your God.
I am Adonai, your God, who brought you out of the land of Egypt to be your God:
I am Adonai your God.

יְיָ אֱלֹֽהֵיכֶם אֱמֶת.

For those who choose: At the word צִיצִת *tzitzit* ("fringes") and at the final word אֱמֶת *emet* ("truth"), one brings the *tzitzit* to one's lips.

וַיֹּ֣אמֶר יְיָ אֶל־מֹשֶׁה *Vayomer Adonai el Moshe . . . Adonai said to Moses . . .* Numbers 15:37–39

לְמַעַן תִּזְכְּרוּ *L'maan tizk'ru . . . Thus you shall remember . . .* Numbers 15:40–41

I If you are wearing a tallit, take the *tzitzit* lightly in your fingers. If not, imagine the tassels in your hand. Feel the freedom of the loose ends. Follow the loose ends up to the knots and the wrappings. Similarly, Judaism binds together the core freedom of our lives and wraps it in commandment and law. In the end, we are free. In the beginning, we are constrained. How is this like your journey from childhood to adulthood?

בָּרְכוּ
יוֹצֵר
אַהֲבָה רַבָּה
שְׁמַע
וְאָהַבְתָּ
לְמַעַן תִּזְכְּרוּ
וַיֹּאמֶר יְיָ
אֱמֶת וְיַצִּיב
מִי־כָמֹכָה

אֱמֶת וְיַצִּיב וְאָהוּב
וְחָבִיב וְנוֹרָא וְאַדִּיר וְטוֹב וְיָפֶה
הַדָּבָר הַזֶּה עָלֵינוּ לְעוֹלָם וָעֶד.
אֱמֶת, אֱלֹהֵי עוֹלָם מַלְכֵּנוּ,
צוּר יַעֲקֹב, מָגֵן יִשְׁעֵנוּ.
לְדֹר וָדֹר הוּא קַיָּם
וּשְׁמוֹ קַיָּם וְכִסְאוֹ נָכוֹן
וּמַלְכוּתוֹ וֶאֱמוּנָתוֹ לָעַד קַיָּמֶת.
וּדְבָרָיו חָיִים וְקַיָּמִים,
נֶאֱמָנִים וְנֶחֱמָדִים
לָעַד וּלְעוֹלְמֵי עוֹלָמִים.
מִמִּצְרַיִם גְּאַלְתָּנוּ, יְיָ אֱלֹהֵינוּ,
וּמִבֵּית עֲבָדִים פְּדִיתָנוּ. עַל זֹאת
שִׁבְּחוּ אֲהוּבִים וְרוֹמְמוּ אֵל,
וְנָתְנוּ יְדִידִים זְמִירוֹת שִׁירוֹת
וְתִשְׁבָּחוֹת, בְּרָכוֹת וְהוֹדָאוֹת
לְמֶלֶךְ אֵל חַי וְקַיָּם. רָם
וְנִשָּׂא, גָּדוֹל וְנוֹרָא, מַשְׁפִּיל
גֵּאִים וּמַגְבִּיהַּ שְׁפָלִים, מוֹצִיא
אֲסִירִים וּפוֹדֶה עֲנָוִים וְעוֹזֵר
דַּלִּים וְעוֹנֶה לְעַמּוֹ בְּעֵת שַׁוְּעָם.
תְּהִלּוֹת לְאֵל עֶלְיוֹן, בָּרוּךְ
הוּא וּמְבֹרָךְ. מֹשֶׁה וּמִרְיָם
וּבְנֵי יִשְׂרָאֵל לְךָ עָנוּ שִׁירָה
בְּשִׂמְחָה רַבָּה וְאָמְרוּ כֻלָּם:

FOR US, this eternal teaching is true and enduring, beloved and precious, awesome, good and beautiful. The God of the universe is truly our Sovereign, the Rock of Jacob, our Protecting Shield. God endures through all generations; God's name persists; God's throne is firm; God's sovereignty and faithfulness last forever. God's words live and endure, faithful and precious for eternity.

From Egypt You redeemed us, freeing us from bondage. For that, Your beloved sang praise, exalting You. Your dear ones offered hymns, songs, praise, blessing, and thanksgiving to You as Sovereign, the living and enduring God. High and exalted, great and awesome, God ever humbles the proud, raises the lowly, frees the imprisoned, redeems the afflicted, helps the oppressed, answering our people when we cry out. Praise to God Most High; blessed is God and deserving of blessing! In great joy, Moses, Miriam and Israel responded with song to You, all of them proclaiming:

Bar'chu
Yotzeir
Ahavah Rabbah
Sh'ma
V'ahavta
L'maan tizk'ru
Vayomer Adonai
Emet v'Yatziv
Mi Chamochah

P The end of this prayer states that God "humbles the proud, raises the lowly, frees the imprisoned, redeems the afflicted, helps the oppressed, answering our people when we cry out." However, we know that there are still afflicted, oppressed, and imprisoned people in the world around us. How do you make sense of this prayer? What does it mean to praise God for these things?

מִי־כָמְכָה בָּאֵלִם, יְיָ!

מִי כָּמְכָה נֶאְדָּר בַּקֹּדֶשׁ,

נוֹרָא תְהִלֹּת, עֹשֵׂה פֶלֶא!

שִׁירָה חֲדָשָׁה שִׁבְּחוּ גְאוּלִים

לְשִׁמְךָ עַל שְׂפַת הַיָּם.

יַחַד כֻּלָּם הוֹדוּ וְהִמְלִיכוּ וְאָמְרוּ:

יְיָ יִמְלֹךְ לְעוֹלָם וָעֶד.

צוּר יִשְׂרָאֵל, קוּמָה בְּעֶזְרַת יִשְׂרָאֵל

וּפְדֵה כִנְאֻמֶךָ יְהוּדָה וְיִשְׂרָאֵל.

גֹּאֲלֵנוּ יְיָ צְבָאוֹת שְׁמוֹ,

קְדוֹשׁ יִשְׂרָאֵל.

בָּרוּךְ אַתָּה, יְיָ, גָּאַל יִשְׂרָאֵל.

בָּרְכוּ

יוֹצֵר

אַהֲבָה רַבָּה

שְׁמַע

וְאָהַבְתָּ

לְמַעַן תִּזְכְּרוּ

וַיֹּאמֶר יְיָ

אֱמֶת וְיַצִּיב

מִי־כָמְכָה

WHO IS LIKE YOU, O God,
among the gods that are worshipped?
Who is like You, majestic in holiness,
awesome in splendor, working wonders?

With new song, inspired,
at the shore of the Sea, the redeemed sang Your praise.
In unison they all offered thanks.
Acknowledging Your Sovereignty, they said:
"Adonai will reign forever!"

Rock of Israel, rise in support of Israel
and redeem Judah and Israel as You promised.
Our Redeemer, *Adonai Tz'vaot* is Your Name.
Blessed are You, Adonai, for redeeming Israel.

בָּרוּךְ אַתָּה, יְיָ, גָּאַל יִשְׂרָאֵל.

For those who choose: When the prayer leader recites the word קוּמָה *kumah* ("rise"), the congregation rises for the עֲמִידָה *Amidah,* the *Standing Prayer.*

מִי־כָמְכָה *Mi chamochah . . . Who is like You . . .* Exodus 15:11

יְיָ יִמְלֹךְ לְעֹלָם וָעֶד *Adonai yimloch l'olam va-ed . . . Adonai will reign forever . . .* Exodus 15:18

I Imagine yourself on the shores of the sea, surrounded by former slaves, your friends and family, now surprisingly free. What is the song that rises from your throat? What thoughts and words want expression? How do you dance in celebration?

אָבוֹת וְאִמָּהוֹת
גְּבוּרוֹת
קְדֻשָּׁה
קְדֻשַּׁת הַיּוֹם
עֲבוֹדָה
הוֹדָאָה
שָׁלוֹם
תְּפִלַּת הַלֵּב

T'FILAH

אֲדֹנָי, שְׂפָתַי תִּפְתָּח,
וּפִי יַגִּיד תְּהִלָּתֶךָ.

ADONAI, open up my lips,
that my mouth may declare Your praise.

For those who choose: Before reciting the תְּפִלָּה *T'filah,* one takes three steps forward.

אֲדֹנָי, שְׂפָתַי תִּפְתָּח *Adonai s'fatai tiftach . . . Adonai, open up my lips . . .* Psalm 51:17

Avot v'Imahot
G'vurot
K'dushah
K'dushat HaYom
Avodah
Hodaah
Shalom
T'filat HaLev

I Feel yourself wrapped in a tallit. Pull it closed around your head. Prepare yourself for prayer in your private space. Feel those around you doing the same. How does the presence of others enhance your personal prayer?

בָּרוּךְ אַתָּה, יְיָ אֱלֹהֵינוּ
וֵאלֹהֵי אֲבוֹתֵינוּ וְאִמּוֹתֵינוּ, אֱלֹהֵי
אַבְרָהָם, אֱלֹהֵי יִצְחָק וֵאלֹהֵי יַעֲקֹב,
אֱלֹהֵי שָׂרָה, אֱלֹהֵי רִבְקָה, אֱלֹהֵי
רָחֵל וֵאלֹהֵי לֵאָה. הָאֵל הַגָּדוֹל
הַגִּבּוֹר וְהַנּוֹרָא, אֵל עֶלְיוֹן, גּוֹמֵל
חֲסָדִים טוֹבִים, וְקוֹנֵה הַכֹּל, וְזוֹכֵר
חַסְדֵי אָבוֹת וְאִמָּהוֹת, וּמֵבִיא גְּאֻלָּה
לִבְנֵי בְנֵיהֶם לְמַעַן שְׁמוֹ בְּאַהֲבָה.

— Shabbat Shuvah*

זָכְרֵנוּ לְחַיִּים,
מֶלֶךְ חָפֵץ בַּחַיִּים,
וְכָתְבֵנוּ בְּסֵפֶר הַחַיִּים,
לְמַעַנְךָ אֱלֹהִים חַיִּים.

מֶלֶךְ עוֹזֵר וּמוֹשִׁיעַ וּמָגֵן.
בָּרוּךְ אַתָּה, יְיָ,
מָגֵן אַבְרָהָם וְעֶזְרַת שָׂרָה.

Blessed are You, Adonai our God,
God of our fathers and mothers,
God of Abraham, God of Isaac, and God of Jacob,
God of Sarah, God of Rebecca, God of Rachel, and God of Leah,
the great, mighty and awesome God, transcendent God
who bestows lovingkindness, creates everything out of love,
remembers the love of our fathers and mothers,
and brings redemption to their children's children for the sake of the Divine Name.

*Shabbat Shuvah — Remember us for life, O Sovereign who delights in life,
and inscribe us in the Book of Life, for Your sake, Living God.

Sovereign, Deliverer, Helper and Shield,
Blessed are You, Adonai, Sarah's Helper, Abraham's Shield.

בָּרוּךְ אַתָּה, יְיָ, מָגֵן אַבְרָהָם וְעֶזְרַת שָׂרָה.

*Shabbat Shuvah: The Shabbat between Rosh HaShanah and Yom Kippur.

For those who choose: At the beginning and end of the blessing, one bends the knees and bows from the waist at the word בָּרוּךְ *Baruch* and stands straight at the word יְיָ *Adonai.*

The content of this prayer has to do with *the merit of our ancestors.* This is traditionally conceived of as a sort of bank account into which the Patriarchs and Matriarchs deposited funds of righteousness that were so great that they covered all future generations. *Judith Z. Abrams*

אַתָּה גִּבּוֹר לְעוֹלָם, אֲדֹנָי,
מְחַיֵּה הַכֹּל (מֵתִים) אַתָּה,
רַב לְהוֹשִׁיעַ.

WINTER* — מַשִּׁיב הָרְוּחַ וּמוֹרִיד הַגֶּשֶׁם.
SUMMER* — מוֹרִיד הַטָּל.

מְכַלְכֵּל חַיִּים בְּחֶסֶד,
מְחַיֵּה הַכֹּל (מֵתִים)
בְּרַחֲמִים רַבִּים, סוֹמֵךְ נוֹפְלִים,
וְרוֹפֵא חוֹלִים, וּמַתִּיר אֲסוּרִים,
וּמְקַיֵּם אֱמוּנָתוֹ לִישֵׁנֵי עָפָר.
מִי כָמְוֹךָ בַּעַל גְּבוּרוֹת
וּמִי דּוֹמֶה לָּךְ, מֶלֶךְ מֵמִית
וּמְחַיֶּה וּמַצְמִיחַ יְשׁוּעָה.

— SHABBAT SHUVAH
מִי כָמְוֹךָ אַב הָרַחֲמִים,
זוֹכֵר יְצוּרָיו לְחַיִּים בְּרַחֲמִים.

וְנֶאֱמָן אַתָּה לְהַחֲיוֹת הַכֹּל (מֵתִים).
בָּרוּךְ אַתָּה, יְיָ, מְחַיֵּה הַכֹּל (הַמֵּתִים).

YOU ARE FOREVER MIGHTY, Adonai; You give life to all (revive the dead).

*WINTER — You cause the wind to shift and rain to fall.

*SUMMER — You rain dew upon us.

You sustain life through love, giving life to all (reviving the dead) through great compassion, supporting the fallen, healing the sick, freeing the captive, keeping faith with those who sleep in the dust. Who is like You, Source of mighty acts? Who resembles You, a Sovereign who takes and gives life, causing deliverance to spring up and faithfully giving life to all (reviving that which is dead)?

SHABBAT SHUVAH — Who is like You, Compassionate God,
who mercifully remembers Your creatures for life?

Blessed are You, Adonai, who gives life to all (revives the dead).
בָּרוּךְ אַתָּה, יְיָ, מְחַיֵּה הַכֹּל (הַמֵּתִים).

The גְּבוּרוֹת *G'vurot* emphasizes God's ability to renew us in the future. The resurrection of the dead, which may be taken literally, is best understood as a powerful metaphor for understanding the miracle of hope. Winter gives way to spring. *Judith Z. Abrams*

*WINTER: *Sh'mini Atzeret / Simchat Torah to Pesach* SUMMER: *Pesach to Sh'mini Atzeret / Simchat Torah*

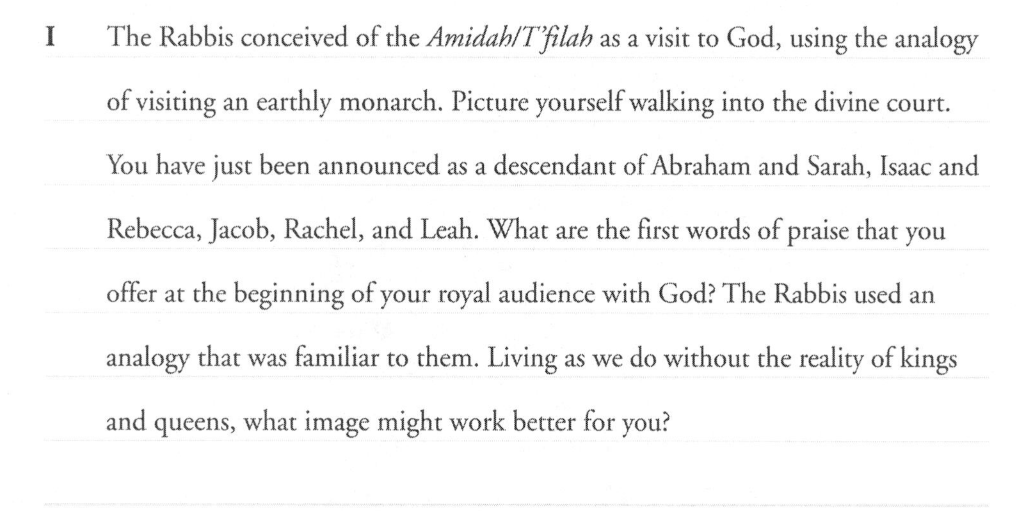

I The Rabbis conceived of the *Amidah/T'filah* as a visit to God, using the analogy of visiting an earthly monarch. Picture yourself walking into the divine court. You have just been announced as a descendant of Abraham and Sarah, Isaac and Rebecca, Jacob, Rachel, and Leah. What are the first words of praise that you offer at the beginning of your royal audience with God? The Rabbis used an analogy that was familiar to them. Living as we do without the reality of kings and queens, what image might work better for you?

מוֹרִיד הַטָּל *Morid hatal . . . You rain dew upon us . . .* A seasonal insertion into the *G'vurot* acknowledges God as the Source of the power of nature. The variations in climate like growth and decay, birth and death, are part of the fixed pattern of the universe created by God. In his prayerbook, *Minhag America*, Isaac Mayer Wise used the prayer for dew and rain as a permanent part of this benediction.

אָבוֹת וְאִמָּהוֹת

גְּבוּרוֹת

קְדֻשָּׁה

קְדֻשַּׁת הַיּוֹם

עֲבוֹדָה

הוֹדָאָה

שָׁלוֹם

תְּפִלַּת הַלֵּב

נְקַדֵּשׁ אֶת שִׁמְךָ בָּעוֹלָם,
כְּשֵׁם שֶׁמַּקְדִּישִׁים אוֹתוֹ בִּשְׁמֵי מָרוֹם,
כַּכָּתוּב עַל יַד נְבִיאֶךָ,
וְקָרָא זֶה אֶל זֶה וְאָמַר:
קָדוֹשׁ, קָדוֹשׁ, קָדוֹשׁ יְיָ צְבָאוֹת,
מְלֹא כָל הָאָרֶץ כְּבוֹדוֹ.
אַדִּיר אַדִּירֵנוּ, יְיָ אֲדוֹנֵנוּ,
מָה אַדִּיר שִׁמְךָ בְּכָל הָאָרֶץ.
בָּרוּךְ כְּבוֹד יְיָ מִמְּקוֹמוֹ.
אֶחָד הוּא אֱלֹהֵינוּ, הוּא אָבִינוּ,
הוּא מַלְכֵּנוּ, הוּא מוֹשִׁיעֵנוּ,
וְהוּא יַשְׁמִיעֵנוּ בְּרַחֲמָיו לְעֵינֵי כָּל חָי.
אֲנִי יְיָ אֱלֹהֵיכֶם.
יִמְלֹךְ יְיָ לְעוֹלָם, אֱלֹהַיִךְ צִיּוֹן
לְדֹר וָדֹר, הַלְלוּיָהּ.

LET US SANCTIFY Your Name on earth, as it is sanctified in the heavens above. As it is written by Your prophet:

Holy, holy, holy is *Adonai Tz'vaot!* God's presence fills the whole earth. Source of our strength, Sovereign One, how majestic is Your presence in all the earth!

Blessed is the presence of God, shining forth from where God dwells. God alone is our God and our Creator, our Ruler and our Helper; and in mercy, God is revealed in the sight of all the living: I am Adonai your God!

Adonai shall reign forever, your God, O Zion,
from generation to generation, Hallelujah!

לְדוֹר וָדוֹר נַגִּיד גָּדְלֶךָ וּלְנֵצַח נְצָחִים
קְדֻשָּׁתְךָ נַקְדִּישׁ, וְשִׁבְחֲךָ, אֱלֹהֵינוּ,
מִפִּינוּ לֹא יָמוּשׁ לְעוֹלָם וָעֶד.*
בָּרוּךְ אַתָּה, יְיָ, הָאֵל הַקָּדוֹשׁ.

TO ALL GENERATIONS we will declare Your greatness, and for all eternity proclaim Your holiness. Your praise, O God, shall never depart from our lips.*
Blessed are You, Adonai, the Holy God.

בָּרוּךְ אַתָּה, יְיָ, הָאֵל הַקָּדוֹשׁ.

*SHABBAT SHUVAH — Blessed are You, Adonai, Holy Sovereign.

בָּרוּךְ אַתָּה, יְיָ, הַמֶּלֶךְ הַקָּדוֹשׁ.

For those who choose: At the words וְקָרָא זֶה *v'kara zeh* one bows to the left and at אֶל זֶה *el zeh* one bows to the right, and at each mention of קָדוֹשׁ *kadosh*, one rises on one's toes.

Avot v'Imahot
G'vurot
K'dushah
K'dushat HaYom
Avodah
Hodaah
Shalom
T'filat HaLev

I This prayer is based on the experience that the prophet Isaiah had when he was transported to heaven to witness the heavenly court praise God (Isaiah 6:3): מְלֹא כָל הָאָרֶץ כְּבוֹדוֹ *M'lo chol haaretz k'vodo,* "God's presence fills the whole earth." If all creation, including us, is part of God's heavenly court, then how do we sing of God's holiness?

Select either Yism'chu or V'shamru

יִשְׂמְחוּ בְמַלְכוּתְךָ
שׁוֹמְרֵי שַׁבָּת וְקוֹרְאֵי עֹנֶג.
עַם מְקַדְּשֵׁי שְׁבִיעִי,
כֻּלָּם יִשְׂבְּעוּ וְיִתְעַנְּגוּ מִטּוּבֶךָ.
וְהַשְּׁבִיעִי רָצִיתָ בּוֹ וְקִדַּשְׁתּוֹ,
חֶמְדַּת יָמִים אוֹתוֹ קָרָאתָ,
זֵכֶר לְמַעֲשֵׂה בְרֵאשִׁית.

גְּבוּרוֹת
קְדֻשָּׁה
קְדֻשַּׁת הַיּוֹם
עֲבוֹדָה
הוֹדָאָה
שָׁלוֹם
תְּפִלַּת הַלֵּב

THOSE WHO KEEP Shabbat by calling it a delight will rejoice in Your realm.
The people that hallows Shabbat will delight in Your goodness.
For, being pleased with the Seventh Day, You hallowed it
as the most precious of days, drawing our attention to the work of Creation.

וְשָׁמְרוּ בְנֵי יִשְׂרָאֵל אֶת־הַשַּׁבָּת,
לַעֲשׂוֹת אֶת־הַשַּׁבָּת לְדֹרֹתָם
בְּרִית עוֹלָם.
בֵּינִי וּבֵין בְּנֵי יִשְׂרָאֵל
אוֹת הִיא לְעֹלָם,
כִּי־שֵׁשֶׁת יָמִים עָשָׂה יְיָ
אֶת־הַשָּׁמַיִם וְאֶת־הָאָרֶץ,
וּבַיּוֹם הַשְּׁבִיעִי שָׁבַת וַיִּנָּפַשׁ.

THE PEOPLE OF ISRAEL shall keep Shabbat,
observing Shabbat throughout the ages as a covenant for all time.
It is a sign for all time between Me and the people of Israel.
For in six days Adonai made heaven and earth,
and on the seventh day God ceased from work and was refreshed.

יִשְׂמְחוּ *Yism'chu* contains twenty-four Hebrew words said to correspond to the twenty-four hours of Shabbat.

וְשָׁמְרוּ בְּנֵי יִשְׂרָאֵל *V'shamru v'nei Yisrael . . . The people of Israel shall keep . . .* Exodus 31:16-17

T Ahad Ha-am, the Cultural Zionist thinker, wrote, "More than the Jewish people has kept Shabbat, Shabbat has kept the Jewish people." What does this mean to you?

אֱלֹהֵינוּ וֵאלֹהֵי
אֲבוֹתֵינוּ וְאִמּוֹתֵינוּ,
רְצֵה בִמְנוּחָתֵנוּ.
קַדְּשֵׁנוּ בְּמִצְוֹתֶיךָ
וְתֵן חֶלְקֵנוּ בְּתוֹרָתֶךָ,
שַׂבְּעֵנוּ מִטּוּבֶךָ,
וְשַׂמְּחֵנוּ בִּישׁוּעָתֶךָ,
וְטַהֵר לִבֵּנוּ לְעָבְדְּךָ בֶּאֱמֶת,
וְהַנְחִילֵנוּ, יְיָ אֱלֹהֵינוּ,
בְּאַהֲבָה וּבְרָצוֹן שַׁבַּת קָדְשֶׁךָ
וְיָנוּחוּ בָהּ יִשְׂרָאֵל, מְקַדְּשֵׁי שְׁמֶךָ.
בָּרוּךְ אַתָּה, יְיָ, מְקַדֵּשׁ הַשַּׁבָּת.

OUR GOD and God of our ancestors,
be pleased with our rest.
Sanctify us with Your mitzvot,
and grant us a share in Your Torah.
Satisfy us with Your goodness
and gladden us with Your salvation.
Purify our hearts to serve You in truth.
In Your gracious love, Adonai our God,
grant us Your holy Shabbat as a heritage.
May Israel who sanctifies Your Name rest on Shabbat.
Blessed are You, Adonai, who sanctifies Shabbat.

בָּרוּךְ אַתָּה, יְיָ, מְקַדֵּשׁ הַשַּׁבָּת.

אָבוֹת וְאִמָּהוֹת

גְּבוּרוֹת

קְדֻשָּׁה

קְדֻשַּׁת הַיּוֹם

עֲבוֹדָה

הוֹדָאָה

שָׁלוֹם

תְּפִלַּת הַלֵּב

רְצֵה, יְיָ אֱלֹהֵינוּ, בְּעַמְּךָ יִשְׂרָאֵל,
וּתְפִלָּתָם בְּאַהֲבָה תְקַבֵּל,
וּתְהִי לְרָצוֹן תָּמִיד
עֲבוֹדַת יִשְׂרָאֵל עַמֶּךָ.
אֵל קָרוֹב לְכָל קֹרְאָיו,
פְּנֵה אֶל עֲבָדֶיךָ וְחָנֵּנוּ,
שְׁפוֹךְ רוּחֲךָ עָלֵינוּ.

FIND FAVOR, Adonai, our God, with Your people Israel
and accept their prayer in love.
May the worship of Your people Israel always be acceptable.
God who is near to all who call, turn lovingly to Your servants.
Pour out Your spirit upon us.

ROSH CHODESH, PESACH, AND SUKKOT

אֱלֹהֵינוּ וֵאלֹהֵי אֲבוֹתֵינוּ וְאִמּוֹתֵינוּ,
יַעֲלֶה וְיָבֹא וְיִזָּכֵר זִכְרוֹנֵנוּ
וְזִכְרוֹן כָּל עַמְּךָ בֵּית יִשְׂרָאֵל לְפָנֶיךָ,
לְטוֹבָה, לְחֵן וּלְחֶסֶד וּלְרַחֲמִים,
לְחַיִּים וּלְשָׁלוֹם, בְּיוֹם

רֹאשׁ הַחֹדֶשׁ הַזֶּה.
חַג הַמַּצּוֹת הַזֶּה.
חַג הַסֻּכּוֹת הַזֶּה.

זָכְרֵנוּ, יְיָ אֱלֹהֵינוּ, בּוֹ לְטוֹבָה. אָמֵן.
וּפָקְדֵנוּ בוֹ לִבְרָכָה. אָמֵן.
וְהוֹשִׁיעֵנוּ בוֹ לְחַיִּים. אָמֵן.

Our God and God of our fathers and mothers, on this
(first day of the new month) — (day of Pesach) — (day of Sukkot)
be mindful of us and all Your people Israel,
for good, for love, for compassion, life and peace.
Remember us for wellbeing. Amen.
Visit us with blessing. Amen. Help us to a fuller life. Amen.

וְתֶחֱזֶינָה עֵינֵינוּ בְּשׁוּבְךָ
לְצִיּוֹן בְּרַחֲמִים.

LET OUR EYES BEHOLD Your loving return to Zion.
Blessed are You, Adonai, whose Presence returns to Zion.

בָּרוּךְ אַתָּה, יְיָ,
הַמַּחֲזִיר שְׁכִינָתוֹ לְצִיּוֹן.

Avot v'Imahot
G'vurot
K'dushah
K'dushat HaYom
Avodah
Hodaah
Shalom
T'filat HaLev

E How often, after an encounter with another, do you wish that you could go back and do things differently? This prayer, offered as we finish the *T'filah,* is a chance for a do-over, a plea that we have prayed correctly, and an apology if we have not. How important is a second chance in prayer as well as in life?

מוֹדִים אֲנַחְנוּ לָךְ, שָׁאַתָּה הוּא
יְיָ אֱלֹהֵינוּ וֵאלֹהֵי אֲבוֹתֵינוּ וְאִמּוֹתֵינוּ
לְעוֹלָם וָעֶד. צוּר חַיֵּינוּ, מָגֵן יִשְׁעֵנוּ,
אַתָּה הוּא לְדוֹר וָדוֹר.

נוֹדֶה לְּךָ וּנְסַפֵּר תְּהִלָּתֶךָ. עַל חַיֵּינוּ
הַמְּסוּרִים בְּיָדֶךָ, וְעַל נִשְׁמוֹתֵינוּ
הַפְּקוּדוֹת לָךְ, וְעַל נִסֶּיךָ שֶׁבְּכָל יוֹם
עִמָּנוּ, וְעַל נִפְלְאוֹתֶיךָ וְטוֹבוֹתֶיךָ
שֶׁבְּכָל עֵת, עֶרֶב וָבֹקֶר וְצָהֳרָיִם.

הַטּוֹב כִּי לֹא כָלוּ רַחֲמֶיךָ, וְהַמְרַחֵם
כִּי לֹא תַמּוּ חֲסָדֶיךָ, מֵעוֹלָם קִוִּינוּ לָךְ.

WE ACKNOWLEDGE with thanks that You are Adonai, our God and the God of our ancestors, forever. You are the Rock of our lives, and the Shield of our salvation in every generation. Let us thank You and praise You — for our lives which are in Your hand, for our souls which are in Your care, for Your miracles that we experience every day and for Your wondrous deeds and favors at every time of day: evening, morning and noon. O Good One, whose mercies never end, O Compassionate One, whose kindness never fails, we forever put our hope in You.

On Chanukah, continue on page 556.

וְעַל כֻּלָּם יִתְבָּרַךְ וְיִתְרוֹמַם שִׁמְךָ,
מַלְכֵּנוּ, תָּמִיד לְעוֹלָם וָעֶד.

SHABBAT SHUVAH — וּכְתוֹב לְחַיִּים
טוֹבִים כָּל בְּנֵי בְרִיתֶךָ.

וְכֹל הַחַיִּים יוֹדוּךָ סֶּלָה,
וִיהַלְלוּ אֶת שִׁמְךָ בֶּאֱמֶת,
הָאֵל יְשׁוּעָתֵנוּ וְעֶזְרָתֵנוּ סֶלָה.
בָּרוּךְ אַתָּה, יְיָ,
הַטּוֹב שִׁמְךָ וּלְךָ נָאֶה לְהוֹדוֹת.

For all these things, O Sovereign, let Your Name be forever praised and blessed.

SHABBAT SHUVAH — Inscribe all the children of Your covenant for a good life.

O God, our Redeemer and Helper, let all who live affirm You and praise Your Name in truth. Blessed are You, Adonai, Your Name is Goodness, and You are worthy of thanksgiving.

בָּרוּךְ אַתָּה, יְיָ, הַטּוֹב שִׁמְךָ וּלְךָ נָאֶה לְהוֹדוֹת.

For those who choose: On page 256, at the word מוֹדִים *Modim,* one bows at the waist. At יְיָ *Adonai,* one stands up straight.

אָבוֹת וְאִמָּהוֹת
גְּבוּרוֹת
קְדֻשָּׁה
קְדֻשַּׁת הַיּוֹם
עֲבוֹדָה
הוֹדָאָה
שָׁלוֹם
תְּפִלַּת הַלֵּב

שִׂים שָׁלוֹם טוֹבָה וּבְרָכָה,
חֵן וָחֶסֶד וְרַחֲמִים,
עָלֵינוּ וְעַל כָּל יִשְׂרָאֵל עַמֶּךָ.
בָּרְכֵנוּ, יוֹצְרֵנוּ, כֻּלָּנוּ כְּאֶחָד
בְּאוֹר פָּנֶיךָ,
כִּי בְאוֹר פָּנֶיךָ נָתַתָּ לָּנוּ,
יְיָ אֱלֹהֵינוּ,
תּוֹרַת חַיִּים וְאַהֲבַת חֶסֶד,
וּצְדָקָה וּבְרָכָה וְרַחֲמִים
וְחַיִּים וְשָׁלוֹם.
וְטוֹב בְּעֵינֶיךָ לְבָרֵךְ אֶת־עַמְּךָ יִשְׂרָאֵל
בְּכָל עֵת וּבְכָל שָׁעָה בִּשְׁלוֹמֶךָ.

SHABBAT SHUVAH — בְּסֵפֶר חַיִּים,
בְּרָכָה וְשָׁלוֹם, וּפַרְנָסָה טוֹבָה,
נִזָּכֵר וְנִכָּתֵב לְפָנֶיךָ,
אֲנַחְנוּ וְכָל עַמְּךָ בֵּית יִשְׂרָאֵל,
לְחַיִּים טוֹבִים וּלְשָׁלוֹם.
בָּרוּךְ אַתָּה, יְיָ, עֹשֵׂה הַשָּׁלוֹם.

בָּרוּךְ אַתָּה, יְיָ,
הַמְבָרֵךְ אֶת עַמּוֹ יִשְׂרָאֵל בַּשָּׁלוֹם.

GRANT PEACE, goodness and blessing, grace, kindness and mercy,
to us and to all Your people Israel.
Bless us, our Creator, all of us together, through the light of Your Presence.
Truly through the light of Your Presence, Adonai our God,
You gave us a Torah of life —
the love of kindness, justice and blessing, mercy, life, and peace.
May You see fit to bless Your people Israel
at all times, at every hour, with Your peace.

SHABBAT SHUVAH — Inscribe us for life, blessing, peace, and prosperity, remembering all Your people Israel for life and peace. Blessed are You, Adonai, Source of peace.
בָּרוּךְ אַתָּה, יְיָ, עֹשֵׂה הַשָּׁלוֹם.

Praised are You, Adonai, who blesses Your people Israel with peace.

בָּרוּךְ אַתָּה, יְיָ, הַמְבָרֵךְ אֶת עַמּוֹ יִשְׂרָאֵל בַּשָּׁלוֹם.

T "Hillel said: Be a disciple of Aaron, loving peace and pursuing peace, loving your fellow creatures and attracting them to the Torah" (*Pirkei Avot* 1:12). How can you apply this teaching to your own life?

אֱלֹהַי, נְצֹר לְשׁוֹנִי מֵרָע וּשְׂפָתַי
מִדַּבֵּר מִרְמָה, וְלִמְקַלְלַי נַפְשִׁי
תִדּוֹם, וְנַפְשִׁי כֶּעָפָר לַכֹּל תִּהְיֶה.
פְּתַח לִבִּי בְּתוֹרָתֶךָ, וּבְמִצְוֹתֶיךָ
תִּרְדּוֹף נַפְשִׁי. וְכֹל הַחוֹשְׁבִים
עָלַי רָעָה, מְהֵרָה הָפֵר עֲצָתָם
וְקַלְקֵל מַחֲשַׁבְתָּם. עֲשֵׂה לְמַעַן
שְׁמֶךָ, עֲשֵׂה לְמַעַן יְמִינֶךָ, עֲשֵׂה
לְמַעַן קְדֻשָּׁתֶךָ, עֲשֵׂה לְמַעַן
תּוֹרָתֶךָ. לְמַעַן יֵחָלְצוּן יְדִידֶיךָ,
הוֹשִׁיעָה יְמִינְךָ וַעֲנֵנִי.

MY GOD, guard my speech from evil and my lips from deception.
Before those who slander me, I will hold my tongue; I will practice humility.
Open my heart to Your Torah, that I may pursue Your mitzvot.
As for all who think evil of me, cancel their designs and frustrate their schemes.
Act for Your own sake, for the sake of Your Power,
for the sake of Your Holiness, for the sake of Your Torah,
so that Your loved ones may be rescued.
Save with Your power, and answer me.

יִהְיוּ לְרָצוֹן אִמְרֵי פִי וְהֶגְיוֹן לִבִּי
לְפָנֶיךָ, יְיָ צוּרִי וְגוֹאֲלִי.

May the words of my mouth and the meditations of my heart
be acceptable to You, Adonai, my Rock and my Redeemer.

עֹשֶׂה שָׁלוֹם בִּמְרוֹמָיו,
הוּא יַעֲשֶׂה שָׁלוֹם עָלֵינוּ,
וְעַל כָּל יִשְׂרָאֵל, וְעַל כָּל יוֹשְׁבֵי תֵבֵל,
וְאִמְרוּ. אָמֵן.

May the One who makes peace in the high heavens
make peace for us, all Israel and all who inhabit the earth. Amen.

Reading of the Torah is on page 362.

אֱלֹהַי, נְצֹר *Elohai, n'tzor . . . My God, guard . . . based on* Psalm 34:14
לְמַעַן יֵחָלְצוּן *L'maan yeichaltzun . . . so that Your loved ones . . .* Psalm 60:7
יִהְיוּ לְרָצוֹן *Yih'yu l'ratzon . . . May the words of my mouth . . .* Psalm 19:15

Avot v'Imahot
G'vurot
K'dushah
K'dushat HaYom
Avodah
Hodaah
Shalom
T'filat HaLev

Note: In order to stay consistent with the pagination of the standard, complete edition of *Mishkan T'filah,* there are several jumps in pagination in this edition.

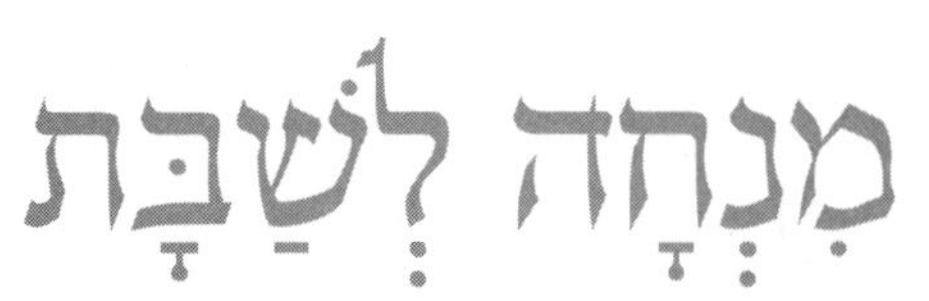

MINCHAH L'SHABBAT — SHABBAT AFTERNOON

אַשְׁרֵי יוֹשְׁבֵי בֵיתֶךָ,
עוֹד יְהַלְלוּךָ סֶּלָה.
אַשְׁרֵי הָעָם שֶׁכָּכָה לּוֹ,
אַשְׁרֵי הָעָם שֶׁיְיָ אֱלֹהָיו.

HAPPY are those who dwell in Your house; they forever praise You! Happy the people who have it so; happy the people whose God is Adonai.

DAVID'S PSALM — PSALM 145:1-21

תְּהִלָּה לְדָוִד
אֲרוֹמִמְךָ אֱלוֹהַי הַמֶּלֶךְ,
וַאֲבָרְכָה שִׁמְךָ לְעוֹלָם וָעֶד.

בְּכָל־יוֹם אֲבָרְכֶךָּ,
וַאֲהַלְלָה שִׁמְךָ לְעוֹלָם וָעֶד.

גָּדוֹל יְיָ וּמְהֻלָּל מְאֹד,
וְלִגְדֻלָּתוֹ אֵין חֵקֶר.

DAVID'S SONG OF PRAISE.
I will extol You, my God and sovereign,
and bless Your name forever and ever.

Every day will I bless You
and praise Your name forever and ever.

Great is Adonai and much acclaimed;
God's greatness cannot be fathomed.

אַשְׁרֵי יוֹשְׁבֵי *Ashrei yoshvei . . . Happy are those who dwell . . .* Psalm 84:5
אַשְׁרֵי הָעָם *Ashrei haam . . . Happy the people . . .* Psalm 144:15

Ashrei

Psalm 145

Chatzi Kaddish

דּוֹר לְדוֹר יְשַׁבַּח מַעֲשֶׂיךָ,
וּגְבוּרֹתֶיךָ יַגִּידוּ.

הֲדַר כְּבוֹד הוֹדֶךָ,
וְדִבְרֵי נִפְלְאֹתֶיךָ אָשִׂיחָה.

וֶעֱזוּז נוֹרְאֹתֶיךָ יֹאמֵרוּ,
וּגְדֻלָּתְךָ אֲסַפְּרֶנָּה.

זֵכֶר רַב־טוּבְךָ יַבִּיעוּ,
וְצִדְקָתְךָ יְרַנֵּנוּ.

חַנּוּן וְרַחוּם יְיָ,
אֶרֶךְ אַפַּיִם וּגְדָל־חָסֶד.

טוֹב יְיָ לַכֹּל,
וְרַחֲמָיו עַל־כָּל־מַעֲשָׂיו.

יוֹדוּךָ יְיָ כָּל־מַעֲשֶׂיךָ,
וַחֲסִידֶיךָ יְבָרְכוּכָה.

One generation shall laud Your works to another
and declare Your mighty acts.

The glorious majesty of Your splendor
and Your wondrous acts will I recite.

They shall talk of the might of Your awesome deeds,
and I will recount Your greatness.

They shall celebrate Your abundant goodness,
and sing joyously of Your beneficence.

Adonai is gracious and compassionate,
slow to anger and abounding in kindness.

Adonai is good to all,
and God's mercy is upon all God's works.

All Your works shall praise You, Adonai,
and Your faithful ones shall bless You.

אַשְׁרֵי
מִזְמוֹר קמ"ה
חֲצִי קַדִּישׁ

כְּבוֹד מַלְכוּתְךָ יֹאמֵרוּ,
וּגְבוּרָתְךָ יְדַבֵּרוּ.

לְהוֹדִיעַ לִבְנֵי הָאָדָם גְּבוּרֹתָיו,
וּכְבוֹד הֲדַר מַלְכוּתוֹ.

מַלְכוּתְךָ מַלְכוּת כָּל־עֹלָמִים,
וּמֶמְשַׁלְתְּךָ בְּכָל־דּוֹר וָדֹר.

סוֹמֵךְ יְיָ לְכָל־הַנֹּפְלִים,
וְזוֹקֵף לְכָל־הַכְּפוּפִים.

עֵינֵי כֹל אֵלֶיךָ יְשַׂבֵּרוּ,
וְאַתָּה נוֹתֵן־לָהֶם אֶת־אָכְלָם בְּעִתּוֹ.

פּוֹתֵחַ אֶת־יָדֶךָ,
וּמַשְׂבִּיעַ לְכָל־חַי רָצוֹן.

They shall talk of the majesty of Your sovereignty,
and speak of Your might,

to make God's mighty acts known among mortals
and the majestic glory of Your sovereignty.

Your sovereignty is eternal;
Your dominion is for all generations.

Adonai supports all who stumble,
and makes all who are bent stand straight.

The eyes of all look to You expectantly,
and You give them their food when it is due.

You give it openhandedly,
feeding every creature to its heart's content.

צַדִּיק יְיָ בְּכָל־דְּרָכָיו,
וְחָסִיד בְּכָל־מַעֲשָׂיו.

קָרוֹב יְיָ לְכָל־קֹרְאָיו,
לְכֹל אֲשֶׁר יִקְרָאֻהוּ בֶאֱמֶת.

רְצוֹן־יְרֵאָיו יַעֲשֶׂה,
וְאֶת־שַׁוְעָתָם יִשְׁמַע וְיוֹשִׁיעֵם.

שׁוֹמֵר יְיָ אֶת־כָּל־אֹהֲבָיו,
וְאֵת כָּל־הָרְשָׁעִים יַשְׁמִיד.

תְּהִלַּת יְיָ יְדַבֶּר־פִּי,
וִיבָרֵךְ כָּל־בָּשָׂר שֵׁם קָדְשׁוֹ
לְעוֹלָם וָעֶד.

וַאֲנַחְנוּ נְבָרֵךְ יָהּ
מֵעַתָּה וְעַד עוֹלָם, הַלְלוּיָהּ.

Adonai is beneficent in all ways
and faithful in all works.

Adonai is near to all who call
to all who call upon God with sincerity.

Adonai fulfills the wishes of those who fear God;
Adonai hears their cry and delivers them.

Adonai watches over all who love God,
but all the wicked God will destroy.

My mouth shall utter the praise of Adonai,
and all creatures shall bless God's holy name forever and ever.

We will bless You God, now and always.
Hallelujah!

וַאֲנַחְנוּ נְבָרֵךְ יָהּ *Vaanachnu n'vareich Yah . . . And all creatures shall bless . . .* Psalm 115:18

אַשְׁרֵי
מִזְמוֹר קמ"ה
חֲצִי קַדִּישׁ

יִתְגַּדַּל וְיִתְקַדַּשׁ שְׁמֵהּ רַבָּא
בְּעָלְמָא דִּי בְרָא כִרְעוּתֵהּ,
וְיַמְלִיךְ מַלְכוּתֵהּ
בְּחַיֵּיכוֹן וּבְיוֹמֵיכוֹן
וּבְחַיֵּי דְכָל בֵּית יִשְׂרָאֵל,
בַּעֲגָלָא וּבִזְמַן קָרִיב,
וְאִמְרוּ: אָמֵן.

יְהֵא שְׁמֵהּ רַבָּא מְבָרַךְ
לְעָלַם וּלְעָלְמֵי עָלְמַיָּא.

יִתְבָּרַךְ וְיִשְׁתַּבַּח וְיִתְפָּאַר
וְיִתְרוֹמַם וְיִתְנַשֵּׂא,
וְיִתְהַדַּר וְיִתְעַלֶּה וְיִתְהַלַּל
שְׁמֵהּ דְּקֻדְשָׁא בְּרִיךְ הוּא,
לְעֵלָּא מִן כָּל בִּרְכָתָא וְשִׁירָתָא,
תֻּשְׁבְּחָתָא וְנֶחֱמָתָא,
דַּאֲמִירָן בְּעָלְמָא, וְאִמְרוּ: אָמֵן.

EXALTED and hallowed be God's great name,
in the world which God created, according to plan.
May God's majesty be revealed in the days of our lifetime
and the life of all Israel —
speedily, imminently.
To which we say: Amen.

Blessed be God's great name to all eternity.

Blessed, praised, honored, exalted,
extolled, glorified, adored, and lauded
be the name of the Holy Blessed One,
beyond all earthly words and songs of blessing, praise, and comfort.
To which we say: Amen.

Reading of the Torah is on page 362.

Ashrei

Psalm 145

Chatzi Kaddish

T'FILAH

אֲדֹנָי, שְׂפָתַי תִּפְתָּח,
וּפִי יַגִּיד תְּהִלָּתֶךָ.

ADONAI, open up my lips
that my mouth may declare Your praise.

For those who choose: Before reciting תְּפִלָּה *T'filah*, one takes three steps forward.

אֲדֹנָי, שְׂפָתַי תִּפְתָּח *Adonai, s'fatai tiftach . . . Adonai, open up my lips . . .* Psalm 51:17

Avot v'Imahot

G'vurot

K'dushah

K'dushat HaYom

Avodah

Hodaah

Shalom

T'filat HaLev

בָּרוּךְ אַתָּה, יְיָ אֱלֹהֵינוּ
וֵאלֹהֵי אֲבוֹתֵינוּ וְאִמּוֹתֵינוּ, אֱלֹהֵי
אַבְרָהָם, אֱלֹהֵי יִצְחָק וֵאלֹהֵי יַעֲקֹב,
אֱלֹהֵי שָׂרָה, אֱלֹהֵי רִבְקָה, אֱלֹהֵי
רָחֵל וֵאלֹהֵי לֵאָה. הָאֵל הַגָּדוֹל
הַגִּבּוֹר וְהַנּוֹרָא, אֵל עֶלְיוֹן, גּוֹמֵל
חֲסָדִים טוֹבִים, וְקוֹנֵה הַכֹּל, וְזוֹכֵר
חַסְדֵי אָבוֹת וְאִמָּהוֹת, וּמֵבִיא גְאֻלָּה
לִבְנֵי בְנֵיהֶם לְמַעַן שְׁמוֹ בְּאַהֲבָה.

— Shabbat Shuvah*

זָכְרֵנוּ לְחַיִּים,
מֶלֶךְ חָפֵץ בַּחַיִּים,
וְכָתְבֵנוּ בְּסֵפֶר הַחַיִּים,
לְמַעַנְךָ אֱלֹהִים חַיִּים.

מֶלֶךְ עוֹזֵר וּמוֹשִׁיעַ וּמָגֵן.
בָּרוּךְ אַתָּה, יְיָ,
מָגֵן אַבְרָהָם וְעֶזְרַת שָׂרָה.

Blessed are You, Adonai our God,
God of our fathers and mothers,
God of Abraham, God of Isaac, and God of Jacob,
God of Sarah, God of Rebecca, God of Rachel, and God of Leah,
the great, mighty and awesome God, transcendent God
who bestows lovingkindness, creates everything out of love,
remembers the love of our fathers and mothers,
and brings redemption to their children's children for the sake of the Divine Name.

*Shabbat Shuvah — Remember us for life, O Sovereign who delights in life,
and inscribe us in the Book of Life, for Your sake, Living God.

Sovereign, Deliverer, Helper and Shield,
Blessed are You, Adonai, Sarah's Helper, Abraham's Shield.

בָּרוּךְ אַתָּה, יְיָ, מָגֵן אַבְרָהָם וְעֶזְרַת שָׂרָה.

*Shabbat Shuvah: The Shabbat between Rosh HaShanah and Yom Kippur.

For those who choose: At the beginning and end of the blessing, one bends the knees and bows from the waist at the word בָּרוּךְ *Baruch* and stands straight at the word יְיָ *Adonai*.

אָבוֹת וְאִמָּהוֹת *Avot v'imahot . . .* As God has been gracious to our forbears, so may we receive divine favor.

אָבוֹת וְאִמָּהוֹת

גְּבוּרוֹת

קְדֻשָּׁה

קְדֻשַּׁת הַיּוֹם

עֲבוֹדָה

הוֹדָאָה

שָׁלוֹם

תְּפִלַּת הַלֵּב

אַתָּה גִּבּוֹר לְעוֹלָם, אֲדֹנָי,
מְחַיֵּה הַכֹּל (מֵתִים) אַתָּה,
רַב לְהוֹשִׁיעַ.

WINTER* — מַשִּׁיב הָרוּחַ
וּמוֹרִיד הַגֶּשֶׁם.

SUMMER* — מוֹרִיד הַטָּל.

מְכַלְכֵּל חַיִּים בְּחֶסֶד,
מְחַיֵּה הַכֹּל (מֵתִים)
בְּרַחֲמִים רַבִּים, סוֹמֵךְ נוֹפְלִים,
וְרוֹפֵא חוֹלִים, וּמַתִּיר אֲסוּרִים,
וּמְקַיֵּם אֱמוּנָתוֹ לִישֵׁנֵי עָפָר.
מִי כָמוֹךָ בַּעַל גְּבוּרוֹת
וּמִי דוֹמֶה לָּךְ, מֶלֶךְ מֵמִית
וּמְחַיֶּה וּמַצְמִיחַ יְשׁוּעָה.

— SHABBAT SHUVAH

מִי כָמוֹךָ אַב הָרַחֲמִים,
זוֹכֵר יְצוּרָיו לְחַיִּים בְּרַחֲמִים.

וְנֶאֱמָן אַתָּה לְהַחֲיוֹת הַכֹּל (מֵתִים).
בָּרוּךְ אַתָּה, יְיָ, מְחַיֵּה הַכֹּל (הַמֵּתִים).

YOU ARE forever mighty, Adonai; You give life to all (revive the dead).

WINTER — You cause the wind to shift and rain to fall.

SUMMER — You rain dew upon us.

You sustain life through love, giving life to all (reviving the dead) through great compassion, supporting the fallen, healing the sick, freeing the captive, keeping faith with those who sleep in the dust. Who is like You, Source of mighty acts? Who resembles You, a Sovereign who takes and gives life, causing deliverance to spring up and faithfully giving life to all (reviving that which is dead)?

SHABBAT SHUVAH — Who is like You, Compassionate God,
who mercifully remembers Your creatures for life?

Blessed are You, Adonai, who gives life to all (who revives the dead).

בָּרוּךְ אַתָּה, יְיָ, מְחַיֵּה הַכֹּל (הַמֵּתִים).

מַשִּׁיב הָרוּחַ / מוֹרִיד הַטָּל *Mashiv haruach / Morid hatal* — with these words, we join our Israeli brothers and sisters in their prayers for seasonal rains in the Land of Israel.

מוֹרִיד הַטָּל *Morid hatal . . . You rain dew upon us . . .* A seasonal insertion into the *G'vurot* acknowledges God as the Source of the power of nature. The variations in climate like growth and decay, birth and death, are part of the fixed pattern of the universe created by God. In his prayerbook, *Minhag America*, Isaac Mayer Wise used the prayer for dew and rain as a permanent part of this benediction.

*WINTER: *Atzeret–Simchat Torah to Pesach.* SUMMER: *Pesach to Sh'mini Atzeret / Simchat Torah.*

Avot v'Imahot
G'vurot
K'dushah
K'dushat HaYom
Avodah
Hodaah
Shalom
T'filat HaLev

The metaphor "reviving the dead" is widely used rabbinically. The Talmud recommends saying בָּרוּךְ אַתָּה יְיָ, מְחַיֵּה הַמֵּתִים *Baruch atah Adonai, m'chayeih hameitim* for greeting a friend after a lapse of seeing the person for twelve months, and for awakening from sleep.

B'rachot 58b, Y. B'rachot 4:2

נְקַדֵּשׁ אֶת שִׁמְךָ בָּעוֹלָם,
כְּשֵׁם שֶׁמַּקְדִּישִׁים אוֹתוֹ בִּשְׁמֵי מָרוֹם,
כַּכָּתוּב עַל יַד נְבִיאֶךָ,
וְקָרָא זֶה אֶל זֶה וְאָמַר:
קָדוֹשׁ, קָדוֹשׁ, קָדוֹשׁ יְיָ צְבָאוֹת,
מְלֹא כָל הָאָרֶץ כְּבוֹדוֹ.
אַדִּיר אַדִּירֵנוּ, יְיָ אֲדוֹנֵנוּ,
מָה אַדִּיר שִׁמְךָ בְּכָל הָאָרֶץ.
בָּרוּךְ כְּבוֹד יְיָ מִמְּקוֹמוֹ.
אֶחָד הוּא אֱלֹהֵינוּ, הוּא אָבִינוּ,
הוּא מַלְכֵּנוּ, הוּא מוֹשִׁיעֵנוּ,
וְהוּא יַשְׁמִיעֵנוּ בְּרַחֲמָיו לְעֵינֵי כָּל חָי.
אֲנִי יְיָ אֱלֹהֵיכֶם.
יִמְלֹךְ יְיָ לְעוֹלָם, אֱלֹהַיִךְ צִיּוֹן
לְדֹר וָדֹר, הַלְלוּיָהּ.

LET US SANCTIFY Your Name on earth, as it is sanctified in the heavens above. As it is written by Your prophet:

Holy, holy, holy is *Adonai Tz'vaot!* God's presence fills the whole earth.
Source of our strength, Sovereign One, how majestic is Your presence in all the earth!
Blessed is the presence of God, shining forth from where God dwells.
God alone is our God and our Creator, our Ruler and our Helper; and in mercy, God is revealed in the sight of all the living: I am Adonai your God!
Adonai shall reign forever, your God, O Zion,
from generation to generation, Hallelujah!

לְדוֹר וָדוֹר נַגִּיד גָּדְלֶךָ וּלְנֵצַח נְצָחִים
קְדֻשָּׁתְךָ נַקְדִּישׁ, וְשִׁבְחֲךָ, אֱלֹהֵינוּ,
מִפִּינוּ לֹא יָמוּשׁ לְעוֹלָם וָעֶד.*
בָּרוּךְ אַתָּה, יְיָ, הָאֵל הַקָּדוֹשׁ.

TO ALL GENERATIONS we will declare Your greatness, and for all eternity proclaim Your holiness. Your praise, O God, shall never depart from our lips.*

Blessed are You, Adonai, the Holy God.

בָּרוּךְ אַתָּה, יְיָ, הָאֵל הַקָּדוֹשׁ.

*SHABBAT SHUVAH — Blessed are You, Adonai, Holy Sovereign.

בָּרוּךְ אַתָּה, יְיָ, הַמֶּלֶךְ הַקָּדוֹשׁ.

For those who choose: At the words וְקָרָא זֶה *v'kara zeh* one bows to the left and at אֶל זֶה *el zeh* one bows to the right, and at each mention of קָדוֹשׁ *kadosh*, one rises on one's toes.

אָבוֹת וְאִמָּהוֹת
גְּבוּרוֹת
קְדֻשָּׁה
קְדֻשַּׁת הַיּוֹם
עֲבוֹדָה
הוֹדָאָה
שָׁלוֹם
תְּפִלַּת הַלֵּב

אַתָּה אֶחָד וְשִׁמְךָ אֶחָד, וּמִי
כְּעַמְּךָ יִשְׂרָאֵל, גּוֹי אֶחָד בָּאָרֶץ,
תִּפְאֶרֶת גְּדֻלָּה וַעֲטֶרֶת יְשׁוּעָה,
יוֹם מְנוּחָה וּקְדֻשָּׁה לְעַמְּךָ נָתָתָּ.
אַבְרָהָם וְשָׂרָה יָגֵילוּ, יִצְחָק וְרִבְקָה
יְרַנֵּנוּ, יַעֲקֹב וְרָחֵל וְלֵאָה וּבְנֵיהֶם
יָנוּחוּ. מְנוּחַת אַהֲבָה וּנְדָבָה, מְנוּחַת
אֱמֶת וֶאֱמוּנָה, מְנוּחַת שָׁלוֹם וְשַׁלְוָה
וְהַשְׁקֵט וָבֶטַח, מְנוּחָה שְׁלֵמָה שֶׁאַתָּה
רוֹצֶה בָּהּ. יַכִּירוּ בָנֶיךָ וְיֵדְעוּ כִּי
מֵאִתְּךָ הִיא מְנוּחָתָם,
וְעַל מְנוּחָתָם יַקְדִּישׁוּ אֶת שְׁמֶךָ.

YOU ARE ONE and Your name is one, and there is none like Your people Israel, a people unique on earth. A garland of glory have You given us, a crown of salvation, a day of rest and holiness. Abraham and Sarah rejoiced in it, Isaac and Rebecca sang, Jacob and Rachel and Leah and their children were refreshed by its rest. A rest of love freely given, a rest of truth and faithfulness, a rest of peace and serenity, tranquility and security, a perfect rest which You so desire. May Your children come to know that this sacred rest links them to You, and through their rest they sanctify Your name.

אֱלֹהֵינוּ וֵאלֹהֵי אֲבוֹתֵינוּ וְאִמּוֹתֵינוּ,
רְצֵה בִמְנוּחָתֵנוּ. קַדְּשֵׁנוּ בְּמִצְוֹתֶיךָ
וְתֵן חֶלְקֵנוּ בְּתוֹרָתֶךָ, שַׂבְּעֵנוּ מִטּוּבֶךָ
וְשַׂמְּחֵנוּ בִּישׁוּעָתֶךָ, וְטַהֵר לִבֵּנוּ
לְעָבְדְּךָ בֶּאֱמֶת. וְהַנְחִילֵנוּ, יְיָ אֱלֹהֵינוּ,
בְּאַהֲבָה וּבְרָצוֹן שַׁבַּת קָדְשֶׁךָ,
וְיָנוּחוּ בָהּ יִשְׂרָאֵל מְקַדְּשֵׁי שְׁמֶךָ.
בָּרוּךְ אַתָּה, יְיָ, מְקַדֵּשׁ הַשַּׁבָּת.

OUR GOD and God of our ancestors, be pleased with our rest. Sanctify us with Your mitzvot, and grant us a share in Your Torah. Satisfy us with Your goodness and gladden us with Your salvation. Purify our hearts to serve You in truth. In Your gracious love, Adonai our God, grant as our heritage Your Holy Shabbat, that Israel who sanctifies Your Name may rest on it. Praise to You, Adonai, who sanctifies Shabbat.

בָּרוּךְ אַתָּה, יְיָ, מְקַדֵּשׁ הַשַּׁבָּת.

Avot v'Imahot
G'vurot
K'dushah
K'dushat HaYom
Avodah
Hodaah
Shalom
T'filat HaLev

E Think back on a project or task that you completed. Did the joy of accomplishment come during the task or after it was completed? How is completion and reflection, sometimes called closure, a necessary part of creation?

אָבוֹת וְאִמָּהוֹת

גְּבוּרוֹת

קְדֻשָּׁה

קְדֻשַּׁת הַיּוֹם

עֲבוֹדָה

הוֹדָאָה

שָׁלוֹם

תְּפִלַּת הַלֵּב

רְצֵה, יְיָ אֱלֹהֵינוּ, בְּעַמְּךָ יִשְׂרָאֵל,
וּתְפִלָּתָם בְּאַהֲבָה תְקַבֵּל,
וּתְהִי לְרָצוֹן תָּמִיד
עֲבוֹדַת יִשְׂרָאֵל עַמֶּךָ.
אֵל קָרוֹב לְכָל קֹרְאָיו,
פְּנֵה אֶל עֲבָדֶיךָ וְחָנֵּנוּ,
שְׁפֹךְ רוּחֲךָ עָלֵינוּ.

FIND FAVOR, Adonai, our God, with Your people Israel
and accept their prayer in love.
May the worship of Your people Israel always be acceptable.
God who is near to all who call, turn lovingly to Your servants.
Pour out Your spirit upon us.

ROSH CHODESH, PESACH, AND SUKKOT

אֱלֹהֵינוּ וֵאלֹהֵי אֲבוֹתֵינוּ וְאִמּוֹתֵינוּ,
יַעֲלֶה וְיָבֹא וְיִזָּכֵר זִכְרוֹנֵנוּ
וְזִכְרוֹן כָּל עַמְּךָ בֵּית יִשְׂרָאֵל לְפָנֶיךָ,
לְטוֹבָה, לְחֵן וּלְחֶסֶד וּלְרַחֲמִים,
לְחַיִּים וּלְשָׁלוֹם, בְּיוֹם

רֹאשׁ הַחֹדֶשׁ הַזֶּה.

חַג הַמַּצּוֹת הַזֶּה.

חַג הַסֻּכּוֹת הַזֶּה.

זָכְרֵנוּ, יְיָ אֱלֹהֵינוּ, בּוֹ לְטוֹבָה. אָמֵן.
וּפָקְדֵנוּ בוֹ לִבְרָכָה. אָמֵן.
וְהוֹשִׁיעֵנוּ בוֹ לְחַיִּים. אָמֵן.

Our God and God of our fathers and mothers, on this
(first day of the new month) — (day of Pesach) — (day of Sukkot)
be mindful of us and all Your people Israel,
for good, for love, for compassion, life and peace.
Remember us for wellbeing. Amen.
Visit us with blessing. Amen. Help us to a fuller life. Amen.

וְתֶחֱזֶינָה עֵינֵינוּ בְּשׁוּבְךָ
לְצִיּוֹן בְּרַחֲמִים.

LET OUR EYES BEHOLD Your loving return to Zion.
Blessed are You, Adonai, whose Presence returns to Zion.

בָּרוּךְ אַתָּה, יְיָ,
הַמַּחֲזִיר שְׁכִינָתוֹ לְצִיּוֹן.

Avot v'Imahot
G'vurot
K'dushah
K'dushat HaYom
Avodah
Hodaah
Shalom
T'filat HaLev

מוֹדִים אֲנַחְנוּ לָךְ, שָׁאַתָּה הוּא
יְיָ אֱלֹהֵינוּ וֵאלֹהֵי אֲבוֹתֵינוּ וְאִמּוֹתֵינוּ
לְעוֹלָם וָעֶד. צוּר חַיֵּינוּ, מָגֵן יִשְׁעֵנוּ,
אַתָּה הוּא לְדוֹר וָדוֹר.

נוֹדֶה לְּךָ וּנְסַפֵּר תְּהִלָּתֶךָ עַל חַיֵּינוּ
הַמְּסוּרִים בְּיָדֶךָ, וְעַל נִשְׁמוֹתֵינוּ
הַפְּקוּדוֹת לָךְ, וְעַל נִסֶּיךָ שֶׁבְּכָל יוֹם
עִמָּנוּ, וְעַל נִפְלְאוֹתֶיךָ וְטוֹבוֹתֶיךָ
שֶׁבְּכָל עֵת, עֶרֶב וָבֹקֶר וְצָהֳרָיִם.

הַטּוֹב כִּי לֹא כָלוּ רַחֲמֶיךָ, וְהַמְרַחֵם
כִּי לֹא תַמּוּ חֲסָדֶיךָ, מֵעוֹלָם קִוִּינוּ לָךְ.

WE ACKNOWLEDGE with thanks that You are Adonai, our God and the God of our ancestors forever. You are the Rock of our lives, and the Shield of our salvation in every generation. Let us thank You and praise You — for our lives which are in Your hand, for our souls which are in Your care, for Your miracles that we experience every day and for Your wondrous deeds and favors at every time of day: evening, morning and noon. O Good One, whose mercies never end, O Compassionate One, whose kindness never fails, we forever put our hope in You.

וְעַל כֻּלָּם יִתְבָּרַךְ וְיִתְרוֹמַם שִׁמְךָ,
מַלְכֵּנוּ, תָּמִיד לְעוֹלָם וָעֶד.

SHABBAT SHUVAH — וּכְתוֹב לְחַיִּים
טוֹבִים כָּל בְּנֵי בְרִיתֶךָ.

וְכֹל הַחַיִּים יוֹדוּךָ סֶּלָה,
וִיהַלְלוּ אֶת שִׁמְךָ בֶּאֱמֶת,
הָאֵל יְשׁוּעָתֵנוּ וְעֶזְרָתֵנוּ סֶלָה.
בָּרוּךְ אַתָּה, יְיָ,
הַטּוֹב שִׁמְךָ וּלְךָ נָאֶה לְהוֹדוֹת.

For all these things, O Sovereign, let Your Name be forever praised and blessed.

SHABBAT SHUVAH — Inscribe all the children of Your covenant for a good life.

O God, our Redeemer and Helper, let all who live affirm You and praise Your Name in truth. Blessed are You, Adonai, Your Name is Goodness, and You are worthy of thanksgiving.

בָּרוּךְ אַתָּה, יְיָ, הַטּוֹב שִׁמְךָ וּלְךָ נָאֶה לְהוֹדוֹת.

For those who choose: On page 356, at the word מוֹדִים *modim*, one bows at the waist. At יְיָ *Adonai* one stands up straight.

שָׁלוֹם רָב עַל יִשְׂרָאֵל עַמְּךָ
תָּשִׂים לְעוֹלָם,
כִּי אַתָּה הוּא מֶלֶךְ אָדוֹן
לְכָל הַשָּׁלוֹם.
וְטוֹב בְּעֵינֶיךָ לְבָרֵךְ
אֶת עַמְּךָ יִשְׂרָאֵל
בְּכָל עֵת וּבְכָל שָׁעָה בִּשְׁלוֹמֶךָ.

SHABBAT SHUVAH — בְּסֵפֶר חַיִּים,
בְּרָכָה וְשָׁלוֹם וּפַרְנָסָה טוֹבָה,
נִזָּכֵר וְנִכָּתֵב לְפָנֶיךָ,
אֲנַחְנוּ וְכָל עַמְּךָ בֵּית יִשְׂרָאֵל,
לְחַיִּים טוֹבִים וּלְשָׁלוֹם.
בָּרוּךְ אַתָּה, יְיָ, עוֹשֵׂה הַשָּׁלוֹם.

בָּרוּךְ אַתָּה, יְיָ,
הַמְבָרֵךְ אֶת עַמּוֹ יִשְׂרָאֵל בַּשָּׁלוֹם.

GRANT ABUNDANT PEACE to Israel Your people forever,
for You are the Sovereign God of all peace.
May it be pleasing to You to bless Your people Israel
in every season and moment with Your peace.

SHABBAT SHUVAH — In the book of life, blessing, peace and prosperity,
may we be remembered and inscribed by You,
we and all Your people Israel for a good life and for peace.
Blessed are You, Adonai, who makes peace.
בָּרוּךְ אַתָּה, יְיָ, עוֹשֵׂה הַשָּׁלוֹם.

Blessed are You, Adonai, who blesses Your people Israel with peace.
בָּרוּךְ אַתָּה, יְיָ, הַמְבָרֵךְ אֶת עַמּוֹ יִשְׂרָאֵל בַּשָּׁלוֹם.

"Seek peace and pursue it." (Psalm 34:15) . . . The midrash observes, we must "seek peace" *in our own place*, and "pursue it" *in every other place. Numbers Rabbah, Chukat 19:27*

Avot v'Imahot
G'vurot
K'dushah
K'dushat HaYom
Avodah
Hodaah
Shalom
T'filat HaLev

T "Peace, peace to the far and the near" (Isaiah 57:19). How do you make peace for yourself? How do you make peace for others? How are these two pursuits different?

אֱלֹהַי, נְצֹר לְשׁוֹנִי מֵרָע וּשְׂפָתַי
מִדַּבֵּר מִרְמָה, וְלִמְקַלְלַי נַפְשִׁי
תִדּוֹם, וְנַפְשִׁי כֶּעָפָר לַכֹּל תִּהְיֶה.
פְּתַח לִבִּי בְּתוֹרָתֶךָ, וּבְמִצְוֹתֶיךָ
תִּרְדּוֹף נַפְשִׁי. וְכָל הַחוֹשְׁבִים
עָלַי רָעָה, מְהֵרָה הָפֵר עֲצָתָם
וְקַלְקֵל מַחֲשַׁבְתָּם. עֲשֵׂה לְמַעַן
שְׁמֶךָ, עֲשֵׂה לְמַעַן יְמִינֶךָ, עֲשֵׂה
לְמַעַן קְדֻשָּׁתֶךָ, עֲשֵׂה לְמַעַן
תוֹרָתֶךָ. לְמַעַן יֵחָלְצוּן יְדִידֶיךָ,
הוֹשִׁיעָה יְמִינְךָ וַעֲנֵנִי.

אָבוֹת וְאִמָּהוֹת
גְּבוּרוֹת
קְדֻשָּׁה
קְדֻשַּׁת הַיּוֹם
עֲבוֹדָה
הוֹדָאָה
שָׁלוֹם
תְּפִלַּת הַלֵּב

MY GOD, guard my speech from evil and my lips from deception.
Before those who slander me, I will hold my tongue; I will practice humility.
Open my heart to Your Torah, that I may pursue Your mitzvot.
As for all who think evil of me, cancel their designs and frustrate their schemes.
Act for Your own sake, for the sake of Your Power,
for the sake of Your Holiness, for the sake of Your Torah,
so that Your loved ones may be rescued.
Save with Your power, and answer me.

יִהְיוּ לְרָצוֹן אִמְרֵי פִי וְהֶגְיוֹן לִבִּי
לְפָנֶיךָ, יְיָ צוּרִי וְגוֹאֲלִי.

May the words of my mouth and the meditations of my heart
be acceptable to You, Adonai, my Rock and my Redeemer.

עֹשֶׂה שָׁלוֹם בִּמְרוֹמָיו,
הוּא יַעֲשֶׂה שָׁלוֹם עָלֵינוּ,
וְעַל כָּל יִשְׂרָאֵל, וְעַל כָּל יוֹשְׁבֵי תֵבֵל,
וְאִמְרוּ. אָמֵן.

May the One who makes peace in the high heavens
make peace for us, all Israel and all who inhabit the earth. Amen.

Aleinu and Kaddish begin on page 586.

אֱלֹהַי, נְצֹר לְשׁוֹנִי *Elohai, n'tzor l'shoni . . . My God, keep my tongue . . . based on* Psalm 34:14

לְמַעַן יֵחָלְצוּן *L'maan yeichaltzun . . . that Your loved ones . . .* Psalm 60:7

יִהְיוּ לְרָצוֹן *Yih'yu l'ratzon . . . May the words of my mouth . . .* Psalm 19:15

סֵדֶר קְרִיאַת הַתּוֹרָה לְשַׁבָּת

SEDER K'RIAT HATORAH L'SHABBAT — READING THE TORAH ON SHABBAT

The Ark is opened. Remove the Torah.

אֵין כָּמוֹךָ בָאֱלֹהִים אֲדֹנָי,
וְאֵין כְּמַעֲשֶׂיךָ.
מַלְכוּתְךָ מַלְכוּת כָּל־עֹלָמִים,
וּמֶמְשַׁלְתְּךָ בְּכָל־דֹּר וָדֹר.
יְיָ מֶלֶךְ, יְיָ מָלָךְ,
יְיָ יִמְלֹךְ לְעֹלָם וָעֶד.
יְיָ עֹז לְעַמּוֹ יִתֵּן,
יְיָ יְבָרֵךְ אֵת עַמּוֹ בַשָּׁלוֹם.

THERE IS NONE LIKE YOU among the gods, Adonai,
and there are no deeds like Yours.
You are Sovereign over all worlds,
and Your dominion is in all generations.
Adonai reigns, Adonai has reigned,
Adonai will reign forever and ever.
Adonai will give strength to our people,
Adonai will bless our people with peace.

אַב הָרַחֲמִים,
הֵיטִיבָה בִרְצוֹנְךָ אֶת צִיּוֹן,
תִּבְנֶה חוֹמוֹת יְרוּשָׁלָיִם.
כִּי בְךָ לְבַד בָּטָחְנוּ,
מֶלֶךְ אֵל רָם וְנִשָּׂא, אֲדוֹן עוֹלָמִים.

Source of mercy: favor Zion with your goodness;
rebuild the walls of Jerusalem.
For in You alone do we trust,
Sovereign, high and exalted, God Eternal.

אֵין כָּמוֹךָ *Ein kamocha . . . There is none like You . . . is an agglomeration of* Psalm 86:8, Psalm 145:13, Psalm 10:16, Psalm 93:1, Exodus 15:18, *and* Psalm 29:11.

הֵיטִיבָה בִרְצוֹנְךָ *heitivah virtzoncha . . . with Your goodness . . .* Psalm 51:20

Kabbalat HaTorah

Hakafah

Birchot HaTorah

Hagbahah

Mi Shebeirach

Birkat HaGomeil

Birchot HaHaftarah

Hachzarat HaTorah

I Walk through the Torah service in your mind: Imagine rising while the Ark is opened. Standing before the open Ark. Taking out the Torah scroll. Walking the scroll around the congregation. Undressing the scroll. Reading from the Torah. Lifting the Torah for the congregation to see. Re-wrapping the scroll. Returning the scroll to the Ark and closing the curtain. Why do you think that our tradition teaches us to show such reverence for the Torah scroll?

קַבָּלַת הַתּוֹרָה
הַקָּפָה
בִּרְכוֹת הַתּוֹרָה
הַגְבָּהָה
מִי שֶׁבֵּרַךְ
בִּרְכַּת הַגּוֹמֵל
בִּרְכוֹת הַהַפְטָרָה
הַחְזָרַת הַתּוֹרָה

כִּי מִצִּיּוֹן תֵּצֵא תוֹרָה,
וּדְבַר־יְיָ מִירוּשָׁלָיִם.

FOR FROM OUT OF ZION will come the Torah,
and the word of Adonai from Jerusalem.

שְׂאוּ שְׁעָרִים רָאשֵׁיכֶם,
וּשְׂאוּ פִּתְחֵי עוֹלָם,
וְיָבוֹא מֶלֶךְ הַכָּבוֹד.
מִי הוּא זֶה מֶלֶךְ הַכָּבוֹד?
יְיָ צְבָאוֹת
הוּא מֶלֶךְ הַכָּבוֹד. סֶלָה.

LIFT UP your heads, O gates! Lift yourselves up, O ancient doors!
Let the Sovereign of glory enter. Who is this Sovereign of glory?
The God of Hosts is the Sovereign of glory!

IN THIS SCROLL IS THE SECRET OF OUR PEOPLE'S LIFE FROM SINAI UNTIL NOW.

Its teaching is love and justice, goodness and hope.
Freedom is its gift to all who treasure it.

שְׂאוּ שְׁעָרִים *S'u sh'arim . . . Lift up . . .* Psalm 24:9-10 is interpreted rabbinically as the psalm recited when the original ark of the covenant was brought to Jerusalem. It is a longstanding Reform innovation in this part of the service, ever since the first German Reform prayer book of 1819 and used therefore, in Reform liturgy as we remove the Torah from the ark.

כִּי מִצִּיּוֹן תֵּצֵא תוֹרָה *Ki mitzion teitzei Torah . . . For from out of Zion . . .* Isaiah 2:3

Kabbalat HaTorah

Hakafah

Birchot HaTorah

Mi Shebeirach

Hagbahah

Birkat HaGomeil

Birchot HaHaftarah

Hachzarat HaTorah

P The Torah service is the center of the prayer service. It is a reenactment of the Revelation at Mount Sinai. What is gained by re-creating that covenantal moment with God each and every time we read Torah? What is the lesson embedded in this structure?

Standing with the Torah, recite these verses:

בָּרוּךְ שֶׁנָּתַן תּוֹרָה
לְעַמּוֹ יִשְׂרָאֵל בִּקְדֻשָּׁתוֹ.

BLESSED IS GOD who in holiness gave the Torah to the people Israel.

שְׁמַע יִשְׂרָאֵל, יְיָ אֱלֹהֵינוּ, יְיָ אֶחָד.

HEAR O ISRAEL, Adonai is our God, Adonai is One.

אֶחָד אֱלֹהֵינוּ, גָּדוֹל אֲדוֹנֵנוּ,
קָדוֹשׁ שְׁמוֹ.

OUR GOD IS ONE, Adonai is great, holy is God's Name.

גַּדְּלוּ לַייָ אִתִּי,
וּנְרוֹמְמָה שְׁמוֹ יַחְדָּו.

EXALT ADONAI with me, let us extol God's Name together.

לְךָ יְיָ הַגְּדֻלָּה וְהַגְּבוּרָה
וְהַתִּפְאֶרֶת וְהַנֵּצַח וְהַהוֹד,
כִּי כֹל בַּשָּׁמַיִם וּבָאָרֶץ.
לְךָ יְיָ הַמַּמְלָכָה
וְהַמִּתְנַשֵּׂא לְכֹל לְרֹאשׁ.

YOURS, ADONAI, is the greatness, might, splendor, triumph, and majesty —
yes, all that is in heaven and on earth.
To You, Adonai, belong sovereignty and preeminence above all.

The Torah is unwrapped.

שְׁמַע יִשְׂרָאֵל *Sh'ma Yisrael . . . Hear O Israel . . .* Deuteronomy 6:4
גַּדְּלוּ לַייָ אִתִּי *Gadlu l'Adonai iti . . . Exalt Adonai with me . . .* Psalm 34:4
לְךָ יְיָ הַגְּדֻלָּה *L'cha Adonai hag'dulah . . . Yours, Adonai, is the greatness . . .* I Chronicles 29:11

HAKAFAH SELECTIONS

רוֹמְמוּ יְיָ אֱלֹהֵינוּ,
וְהִשְׁתַּחֲווּ לְהַר קָדְשׁוֹ,
כִּי קָדוֹשׁ יְיָ אֱלֹהֵינוּ.

EXALT ADONAI our God and bow down toward God's holy mountain,
for Adonai our God is holy.

עַל שְׁלֹשָׁה דְּבָרִים הָעוֹלָם עוֹמֵד:
עַל הַתּוֹרָה וְעַל הָעֲבוֹדָה
וְעַל גְּמִילוּת חֲסָדִים.

THE WORLD is sustained by three things: Torah, worship and loving deeds.

לֹא־יִשָּׂא גוֹי אֶל־גּוֹי חֶרֶב
וְלֹא־יִלְמְדוּ עוֹד מִלְחָמָה.

NATION SHALL not lift up sword against nation;
neither shall they learn war anymore.

הַלְלוּ . . .
כֹּל הַנְּשָׁמָה תְּהַלֵּל יָהּ,
הַלְלוּ, הַלְלוּ־יָהּ.

LET all that breathes praise God. Hallelujah!

רוֹמְמוּ יְיָ *Rom'mu Adonai . . . Exalt Adonai . . .* Psalm 99:9

עַל שְׁלֹשָׁה דְּבָרִים *Al sh'loshah d'varim . . . The world is sustained by three things . . . Pirkei Avot* 1:2

לֹא־יִשָּׂא גוֹי *Lo yisa goy . . . Nation shall not lift up . . .* Isaiah 2:4

כֹּל הַנְּשָׁמָה *Kol han'shamah . . . Let all that breathes . . .* Psalm 150:6

הָבוּ גֹדֶל לֵאלֹהֵינוּ,
וּתְנוּ כָבוֹד לַתּוֹרָה.

LET US DECLARE the greatness of our God and give honor to the Torah.

ONE WHO MAKES AN ALIYAH MIGHT OFFER:

MAY GOD be with you! — יְיָ עִמָּכֶם.

Congregation responds:

MAY GOD bless you! — יְבָרֶכְךָ יְיָ.

BLESSING BEFORE THE READING OF THE TORAH

בָּרְכוּ אֶת יְיָ הַמְבֹרָךְ.
בָּרוּךְ יְיָ הַמְבֹרָךְ לְעוֹלָם וָעֶד.
בָּרוּךְ אַתָּה, יְיָ
אֱלֹהֵינוּ, מֶלֶךְ הָעוֹלָם,
אֲשֶׁר בָּחַר בָּנוּ מִכָּל הָעַמִּים,
וְנָתַן לָנוּ אֶת תּוֹרָתוֹ.
בָּרוּךְ אַתָּה, יְיָ, נוֹתֵן הַתּוֹרָה.

BLESS ADONAI who is blessed.
Blessed is Adonai who is blessed now and forever.
Blessed are You, Adonai our God, Sovereign of the universe, who has chosen us from among the peoples, and given us the Torah. Blessed are You, Adonai, who gives the Torah.

BLESSING AFTER THE READING OF THE TORAH

בָּרוּךְ אַתָּה, יְיָ
אֱלֹהֵינוּ, מֶלֶךְ הָעוֹלָם,
אֲשֶׁר נָתַן לָנוּ תּוֹרַת אֱמֶת,
וְחַיֵּי עוֹלָם נָטַע בְּתוֹכֵנוּ.
בָּרוּךְ אַתָּה, יְיָ, נוֹתֵן הַתּוֹרָה.

BLESSED ARE YOU, Adonai our God, Sovereign of the universe,
who has given us a Torah of truth, implanting within us eternal life.
Blessed are You, Adonai, who gives the Torah.

I To bless or read from the Torah, we make עֲלִיָּה *aliyah,* we "go up." What does it mean to "go up" to the Torah? What do we bring back down from that experience?

קַבָּלַת הַתּוֹרָה

הַקָּפָה

בִּרְכוֹת הַתּוֹרָה

מִי שֶׁבֵּרַךְ

הַגְבָּהָה

בִּרְכַּת הַגּוֹמֵל

בִּרְכוֹת הַהַפְטָרָה

הַחְזָרַת הַתּוֹרָה

MI SHEBEIRACH FOR ALIYAH

מִי שֶׁבֵּרַךְ אֲבוֹתֵינוּ וְאִמּוֹתֵינוּ,
אַבְרָהָם יִצְחָק וְיַעֲקֹב,
שָׂרָה, רִבְקָה, רָחֵל וְלֵאָה,
הוּא יְבָרֵךְ אֶת *[name]* בֶּן/בַּת *[parents]*
בַּעֲבוּר שֶׁעָלָה/שֶׁעָלְתָה
לִכְבוֹד הַמָּקוֹם, וְלִכְבוֹד הַתּוֹרָה.
בִּשְׂכַר זֶה הַקָּדוֹשׁ בָּרוּךְ הוּא
יִשְׁמְרֵהוּ/יִשְׁמְרֶהָ
וְיַצִּילֵהוּ/וְיַצִּילֶהָ
מִכָּל צָרָה וְצוּקָה וּמִכָּל נֶגַע וּמַחֲלָה,
וְיִשְׁלַח בְּרָכָה וְהַצְלָחָה
בְּכָל מַעֲשֵׂה יָדָיו/יָדֶיהָ,
עִם כָּל יִשְׂרָאֵל. וְנֹאמַר: אָמֵן.

MAY THE ONE WHO BLESSED our ancestors, Abraham, Isaac and Jacob, Sarah, Rebecca, Rachel and Leah, bless *[name]* son/daughter of *[parents]*, since he/she has come up to the Torah in honor of God and Torah. May he/she merit from the Holy One of Blessing protection, rescue from any trouble or distress, and from any illness, minor or serious; may God send blessing and success in his/her every endeavor, together with all Israel, and let us say, Amen.

HAGBAHAH UG'LILAH— הַגְבָּהָה וּגְלִילָה
The Torah is raised, rolled and wrapped.

וְזֹאת הַתּוֹרָה אֲשֶׁר שָׂם מֹשֶׁה
לִפְנֵי בְּנֵי יִשְׂרָאֵל,
עַל־פִּי יְיָ בְּיַד־מֹשֶׁה.

THIS IS THE TORAH which Moses placed
before the people of Israel,
God's word through the hand of Moses.

Prayers of Our Community begin on page 376.
Shabbat Minchah T'filah is on pages 344–345.

וְזֹאת הַתּוֹרָה *V'zot haTorah . . . This is the Torah . . .* Deuteronomy 4:44

עַל־פִּי יְיָ *al pi Adonai . . . God's word . . .* Numbers 9:23

PRAYERS FOR HEALING

מִי שֶׁבֵּרַךְ אֲבוֹתֵינוּ וְאִמּוֹתֵינוּ,
אַבְרָהָם, יִצְחָק וְיַעֲקֹב, שָׂרָה, רִבְקָה,
רָחֵל וְלֵאָה, הוּא יְבָרֵךְ אֶת הַחוֹלִים
[*names*]. הַקָּדוֹשׁ בָּרוּךְ הוּא יִמָּלֵא
רַחֲמִים עֲלֵיהֶם, לְהַחֲלִימָם וּלְרַפֹּאתָם
וּלְהַחֲזִיקָם, וְיִשְׁלַח לָהֶם מְהֵרָה
רְפוּאָה, רְפוּאָה שְׁלֵמָה מִן הַשָּׁמַיִם,
רְפוּאַת הַנֶּפֶשׁ וּרְפוּאַת הַגּוּף, הַשְׁתָּא
בַּעֲגָלָא וּבִזְמַן קָרִיב. וְנֹאמַר: אָמֵן.

MAY THE ONE who blessed our ancestors, Abraham, Isaac and Jacob, Sarah, Rebecca, Rachel and Leah, bless and heal those who are ill [*names*]. May the Blessed Holy One be filled with compassion for their health to be restored and their strength to be revived. May God swiftly send them a complete renewal of body and spirit, and let us say, Amen.

מִי שֶׁבֵּרַךְ אֲבוֹתֵינוּ
מְקוֹר הַבְּרָכָה לְאִמּוֹתֵינוּ.

May the Source of strength who blessed the ones before us
help us find the courage to make our lives a blessing and let us say, Amen.

מִי שֶׁבֵּרַךְ אִמּוֹתֵינוּ
מְקוֹר הַבְּרָכָה לַאֲבוֹתֵינוּ.

Bless those in need of healing with *r'fuah sh'leimah,*
the renewal of body, the renewal of spirit, and let us say, Amen.

BIRKAT HAGOMEIL — בִּרְכַּת הַגּוֹמֵל — THANKSGIVING BLESSING

Individual recites:

בָּרוּךְ אַתָּה, יְיָ אֱלֹהֵינוּ,
מֶלֶךְ הָעוֹלָם, שֶׁגְּמָלַנִי כָּל טוֹב.

BLESSED ARE YOU, Adonai our God, Sovereign of the universe,
who has bestowed every goodness upon us.

Congregation responds:

אָמֵן. מִי שֶׁגְּמָלְכֶם כָּל טוֹב,
הוּא יִגְמָלְכֶם כָּל טוֹב סֶלָה.

Amen. May the One who has bestowed goodness upon us
continue to bestow every goodness upon us forever.

Shabbat Minchah T'filah is on pages 344–345.

בִּרְכַּת הַגּוֹמֵל *Birkat HaGomeil*— may be recited by one who has survived a life-challenging situation.

BLESSING BEFORE THE HAFTARAH

בָּרוּךְ אַתָּה, יְיָ
אֱלֹהֵינוּ, מֶלֶךְ הָעוֹלָם,
אֲשֶׁר בָּחַר בִּנְבִיאִים טוֹבִים,
וְרָצָה בְדִבְרֵיהֶם
הַנֶּאֱמָרִים בֶּאֱמֶת.
בָּרוּךְ אַתָּה, יְיָ, הַבּוֹחֵר בַּתּוֹרָה
וּבְמֹשֶׁה עַבְדּוֹ, וּבְיִשְׂרָאֵל עַמּוֹ,
וּבִנְבִיאֵי הָאֱמֶת וָצֶדֶק.

PRAISE TO YOU, Adonai our God, Sovereign of the universe,
who has chosen faithful prophets to speak words of truth.
Praise to You, Adonai, for the revelation of Torah, for Your servant Moses,
for Your people Israel and for prophets of truth and righteousness.

BLESSING AFTER THE HAFTARAH

בָּרוּךְ אַתָּה, יְיָ אֱלֹהֵינוּ, מֶלֶךְ
הָעוֹלָם, צוּר כָּל הָעוֹלָמִים, צַדִּיק
בְּכָל הַדּוֹרוֹת, הָאֵל הַנֶּאֱמָן,
הָאוֹמֵר וְעֹשֶׂה, הַמְדַבֵּר וּמְקַיֵּם,
שֶׁכָּל דְּבָרָיו אֱמֶת וָצֶדֶק.

עַל הַתּוֹרָה, וְעַל הָעֲבוֹדָה, וְעַל
הַנְּבִיאִים, וְעַל יוֹם הַשַּׁבָּת הַזֶּה,
שֶׁנָּתַתָּ לָּנוּ יְיָ אֱלֹהֵינוּ, לִקְדֻשָּׁה
וְלִמְנוּחָה, לְכָבוֹד וּלְתִפְאָרֶת.

עַל הַכֹּל יְיָ אֱלֹהֵינוּ, אֲנַחְנוּ מוֹדִים
לָךְ, וּמְבָרְכִים אוֹתָךְ, יִתְבָּרַךְ שִׁמְךָ
בְּפִי כָל חַי תָּמִיד לְעוֹלָם וָעֶד.
בָּרוּךְ אַתָּה, יְיָ, מְקַדֵּשׁ הַשַּׁבָּת.

PRAISE TO YOU, Adonai our God, Sovereign of the universe,
Rock of all creation, Righteous One of all generations,
the faithful God whose word is deed, whose every command is just and true.
For the Torah, for the privilege of worship, for the prophets,
and for this Shabbat that You, Adonai our God, have given us
for holiness and rest, for honor and glory: we thank and bless You.
May Your name be blessed for ever by every living being.
Praise to You, Adonai, for the Sabbath and its holiness.

T The prayer before the reading of the haftarah thanks and praises God for providing "prophets of truth and righteousness." What is the role of prophets in a society? Who do you think are contemporary prophets?

יְהַלְלוּ אֶת שֵׁם יְיָ,
כִּי נִשְׂגָּב שְׁמוֹ לְבַדּוֹ.

LET US PRAISE the Name of Adonai,
for God's Name alone is exalted!

הוֹדוֹ עַל אֶרֶץ וְשָׁמָיִם.
וַיָּרֶם קֶרֶן לְעַמּוֹ,
תְּהִלָּה לְכָל־חֲסִידָיו,
לִבְנֵי יִשְׂרָאֵל עַם־קְרֹבוֹ.
הַלְלוּ־יָהּ!

GOD'S MAJESTY is above the earth and heaven; and God is the strength of our people, making God's faithful ones, Israel, a people close to the Eternal. Halleluyah!

The Torah is returned to the Ark.

כִּי לֶקַח טוֹב נָתַתִּי לָכֶם,
תּוֹרָתִי אַל תַּעֲזֹבוּ.

עֵץ חַיִּים הִיא לַמַּחֲזִיקִים בָּהּ,
וְתֹמְכֶיהָ מְאֻשָּׁר.
דְּרָכֶיהָ דַרְכֵי נֹעַם,
וְכָל נְתִיבוֹתֶיהָ שָׁלוֹם.

הֲשִׁיבֵנוּ יְיָ אֵלֶיךָ וְנָשׁוּבָה,
חַדֵּשׁ יָמֵינוּ כְּקֶדֶם.

FOR I HAVE GIVEN YOU good instruction; do not abandon My Torah.

IT IS A TREE OF LIFE for those who hold fast to it, and all its supporters are happy.
Its ways are ways of pleasantness and all its paths are peace.
Return us to You, Adonai, and we will return; renew our days as of old.

יְהַלְלוּ . . . הוֹדוֹ *Y'hal'lu . . . Hodo . . . Let us praise . . . God's majesty* Psalm 148:13-14

כִּי לֶקַח טוֹב *Ki lekach tov . . . For I have given you . . . is an agglomeration of* Proverbs 4:2, Proverbs 3:18, Proverbs 3:17, *and* Lamentations 5:21

Kabbalat HaTorah

Hakafah

Birchot HaTorah

Mi Shebeirach

Hagbahah

Birkat HaGomeil

Birchot HaHaftarah

Hachzarat HaTorah

COMMUNITY

FOR OUR CONGREGATION

SOURCE of all being,
may the children of this community learn these passions from us:
love of Torah, devotion in prayer, and support of the needy.
May we guide with integrity, and may our leadership be in Your service.
May those who teach and nourish us be blessed with satisfaction,
and may we appreciate their time and their devotion.
Bless us with the fruits of wisdom and understanding,
and may our efforts bring fulfillment and joy.

בָּרוּךְ אַתָּה, יְיָ, שֶׁאוֹתְךָ לְבַדְּךָ בְּיִרְאָה נַעֲבוֹד.

FOR OUR COUNTRY

THUS SAYS ADONAI, This is what I desire:
to unlock the fetters of wickedness, and untie the cords of lawlessness;
to let the oppressed go free, to break off every yoke.
Share your bread with the hungry, and take the wretched poor into your home.
When you see the naked, give clothing, and do not ignore your own kin.

If you banish the yoke from your midst, the menacing hand, the evil speech;
if you offer compassion to the hungry and satisfy the famished creature —
then your light shall shine in darkness.

O GUARDIAN of life and liberty,
may our nation always merit Your protection.
Teach us to give thanks for what we have
by sharing it with those who are in need.
Keep our eyes open to the wonders of creation,
and alert to the care of the earth.
May we never be lazy in the work of peace;
may we honor those who have died in defense of our ideals.
Grant our leaders wisdom and forebearance.
May they govern with justice and compassion
Help us all to appreciate one another,
and to respect the many ways that we may serve You.
May our homes be safe from affliction and strife,
and our country be sound in body and spirit.
Amen.

Thus says Adonai . . .   Selected verses from Isaiah 58

PRAYERS OF OUR

FOR THE STATE OF ISRAEL

שַׁאֲלוּ שְׁלוֹם יְרוּשָׁלָיִם,
יִשְׁלָיוּ אֹהֲבָיִךְ.

PRAY for the peace of Jerusalem;
may those who love you prosper.

אָבִינוּ שֶׁבַּשָּׁמַיִם,
צוּר יִשְׂרָאֵל וְגוֹאֲלוֹ,
בָּרֵךְ אֶת מְדִינַת יִשְׂרָאֵל,
רֵאשִׁית צְמִיחַת גְּאֻלָּתֵנוּ.
הָגֵן עָלֶיהָ בְּאֶבְרַת חַסְדֶּךָ,
וּפְרוֹס עָלֶיהָ סֻכַּת שְׁלוֹמֶךָ.
וּשְׁלַח אוֹרְךָ וַאֲמִתְּךָ לְרָאשֶׁיהָ,
שָׂרֶיהָ וְיוֹעֲצֶיהָ,
וְתַקְּנֵם בְּעֵצָה טוֹבָה מִלְּפָנֶיךָ.
וְנָתַתָּ שָׁלוֹם בָּאָרֶץ,
וְשִׂמְחַת עוֹלָם לְיוֹשְׁבֶיהָ.
וְנֹאמַר: אָמֵן.

O HEAVENLY ONE, Protector and Redeemer of Israel,
bless the State of Israel which marks the dawning of hope for all who seek peace.
Shield it beneath the wings of Your love; spread over it the canopy of Your peace;
send Your light and truth to all who lead and advise,
guiding them with Your good counsel.
Establish peace in the land and fullness of joy for all who dwell there.
Amen.

שַׁאֲלוּ שְׁלוֹם יְרוּשָׁלָיִם *Shaalu sh'lom Y'rushalayim . . . Pray for the peace of Jerusalem . . .* Psalm 122:6

COMMUNITY

T'FILAT HADERECH — תְּפִלַּת הַדֶּרֶךְ — UPON SETTING FORTH ON A JOURNEY

יְהִי רָצוֹן מִלְּפָנֶיךָ יְיָ אֱלֹהֵינוּ
וֵאלֹהֵי אֲבוֹתֵינוּ וְאִמּוֹתֵינוּ,
שֶׁתּוֹלִיכֵנוּ לְשָׁלוֹם
וְתַעַזְרֵנוּ לְהַגִּיעַ לִמְחוֹז חֶפְצֵנוּ
לְחַיִּים וּלְשִׂמְחָה וּלְשָׁלוֹם.
וּשְׁמוֹר צֵאתֵנוּ וּבוֹאֵנוּ
וְתַצִּילֵנוּ מִכָּל צָרָה
וְתִשְׁלַח בְּרָכָה בְּכָל מַעֲשֵׂי יָדֵינוּ,
וּמַעֲשֵׂינוּ יְכַבְּדוּ אֶת שְׁמֶךָ.
בָּרוּךְ אַתָּה, יְיָ, שׁוֹמֵר יִשְׂרָאֵל לָעַד.

MAY IT BE YOUR WILL, our God and God of our ancestors,
that You lead us in peace and help us reach our destination
safely, joyfully and peacefully.
May You protect us on our leaving and on our return,
and rescue us from any harm,
and may You bless the work of our hands,
and may our deeds merit honor for You.
Praise to You, Adonai, Protector of Israel.

בָּרוּךְ אַתָּה, יְיָ, שׁוֹמֵר יִשְׂרָאֵל לָעַד.

PRAYERS OF OUR

L'ROSH CHODESH — לְרֹאשׁ חֹדֶשׁ — FOR THE NEW MONTH

יְהִי רָצוֹן מִלְּפָנֶיךָ,
יְיָ אֱלֹהֵינוּ
וֵאלֹהֵי אֲבוֹתֵינוּ וְאִמּוֹתֵינוּ,
שֶׁתְּחַדֵּשׁ עָלֵינוּ אֶת הַחֹדֶשׁ
הַבָּא (הַזֶּה) לְטוֹבָה וְלִבְרָכָה.
וְתִתֶּן לָנוּ חַיִּים אֲרֻכִּים,
חַיִּים שֶׁל שָׁלוֹם,
חַיִּים שֶׁל פַּרְנָסָה,
חַיִּים שֶׁתְּהֵא בָנוּ
אַהֲבַת תּוֹרָה וְיִרְאַת שָׁמַיִם,
חַיִּים שֶׁיִּמָּלְאוּ מִשְׁאֲלוֹת
לִבֵּנוּ לְטוֹבָה. אָמֵן.

OUR GOD and God of our ancestors,
may the new month bring us goodness and blessing.
May we have long life, peace, prosperity,
a life exalted by love of Torah and reverence for the divine;
a life in which the longings of our hearts are fulfilled for good.

רֹאשׁ חֹדֶשׁ *[name of month]* יִהְיֶה בְּיוֹם *[day]* / הוּא הַיּוֹם

THE NEW MONTH of ______ will begin on ______ / begins today.

The custom of announcing the beginning of the new Hebrew month in the synagogue dates to the geonic period (circa 9th century). This prayer is recited on the Shabbat preceding the beginning of the new Hebrew month, with the exception of the month of Tishrei which always coincides with Rosh HaShanah. This Shabbat is called Shabbat M'vorchim. We pray for blessing in the month ahead.

COMMUNITY

FOR A BAR AND BAT MITZVAH

INTO OUR HANDS, O God, You have placed Your Torah, to be held high by parents and children, and taught by one generation to the next. Whatever has befallen us, our people have remained steadfast in loyalty to the Torah. It was carried in the arms of parents that their children might not be deprived of their birthright.

And now, we pray that you, [*name*], may always be worthy of this inheritance. Take its teaching into your heart, and in turn pass it on to your children and those who come after you. May you be a faithful Jew, searching for wisdom and truth, working for justice and peace.

May the One who has always been our Guide inspire you to bring honor to our family and to the House of Israel.

בָּרוּךְ אַתָּה, יְיָ, שֶׁנָּתַן לִי אֶת הַזְּכוּת וְאֶת הַכָּבוֹד לָתֵת לְךָ/לָךְ תּוֹרָה.

Blessed is Adonai our God, who gives me the honor and privilege
of entrusting you with Torah.

OUR HEARTS are one on this joyous day
as you commit yourself to a life of Torah:
a life, we pray, filled with
wisdom, caring and right action.

We pray that you will grow each day
in compassion for the needy,
in concern for the stranger,
in love of all people.

May the One who blessed our ancestors,
Abraham and Sarah, Isaac and Rebecca,
Jacob and Rachel and Leah,
bless you on your becoming a Bar/t Mitzvah.

May you grow with strength and courage,
with vision and sensitivity.
And may you always be certain of our love.

Amen.

T'filot K'hilah

T'filat Haderech

Rosh Chodesh

Blessing for Bar/Bat Mitzvah

PRAYERS OF OUR

E Individual communities write their own prayers. What is your prayer or blessing for the community with which you are worshipping?

Note: In order to stay consistent with the pagination of the standard, complete edition of *Mishkan T'filah,* there are several jumps in pagination in this edition.

עָלֵינוּ וְקַדִּישׁ יָתוֹם

ALEINU V'KADDISH YATOM

Aleinu and Mourner's Kaddish

עָלֵינוּ

ALEINU

Select one of the passages on this page.

עָלֵינוּ לְשַׁבֵּחַ לַאֲדוֹן הַכֹּל,
לָתֵת גְּדֻלָּה לְיוֹצֵר בְּרֵאשִׁית,
שֶׁהוּא נוֹטֶה שָׁמַיִם וְיֹסֵד אָרֶץ,
וּמוֹשַׁב יְקָרוֹ בַּשָּׁמַיִם מִמַּעַל,
וּשְׁכִינַת עֻזּוֹ בְּגָבְהֵי מְרוֹמִים,
הוּא אֱלֹהֵינוּ אֵין עוֹד.
וַאֲנַחְנוּ כּוֹרְעִים
וּמִשְׁתַּחֲוִים וּמוֹדִים,
לִפְנֵי מֶלֶךְ מַלְכֵי הַמְּלָכִים
הַקָּדוֹשׁ בָּרוּךְ הוּא.

LET US NOW PRAISE the Sovereign of the universe, and proclaim the greatness of the Creator who spread out the heavens and established the earth, whose glory is revealed in the heavens above and whose greatness is manifest throughout the world. You are our God; there is none else. Therefore we bow in awe and thanksgiving before the One who is Sovereign over all, the Holy and Blessed One.

Continue on page 588.

עָלֵינוּ לְשַׁבֵּחַ לַאֲדוֹן הַכֹּל,
לָתֵת גְּדֻלָּה לְיוֹצֵר בְּרֵאשִׁית,
שֶׁלֹּא עָשָׂנוּ כְּגוֹיֵי הָאֲרָצוֹת,
וְלֹא שָׂמָנוּ כְּמִשְׁפְּחוֹת הָאֲדָמָה.
שֶׁלֹּא שָׂם חֶלְקֵנוּ כָּהֶם,
וְגֹרָלֵנוּ כְּכָל־הֲמוֹנָם.
וַאֲנַחְנוּ כּוֹרְעִים
וּמִשְׁתַּחֲוִים וּמוֹדִים,
לִפְנֵי מֶלֶךְ מַלְכֵי הַמְּלָכִים
הַקָּדוֹשׁ בָּרוּךְ הוּא.

LET US NOW PRAISE the Sovereign of the universe, and proclaim the greatness of the Creator who has set us apart from the other families of the earth, giving us a destiny unique among the nations. We bend the knee and bow, acknowledging the supreme Sovereign, the Holy One of Blessing.

Continue on page 588.

For those who choose: At the word כּוֹרְעִים *kor'im,* one bends the knees; at וּמִשְׁתַּחֲוִים *umishtachavim*, one bows at the waist; and at לִפְנֵי מֶלֶךְ *lifnei Melech*, one stands straight.

Aleinu

I During the *Aleinu,* as you bend your knees and bow, pay attention to the pull of gravity downward. As you rise up again, feel how your body pushes against the earth. What does it mean to you to bow and feel the pull of the world in front of God?

שֶׁהוּא נוֹטֶה שָׁמַיִם וְיֹסֵד אָרֶץ,
וּמוֹשַׁב יְקָרוֹ בַּשָּׁמַיִם מִמַּעַל
וּשְׁכִינַת עֻזּוֹ בְּגָבְהֵי מְרוֹמִים.
הוּא אֱלֹהֵינוּ אֵין עוֹד,
אֱמֶת מַלְכֵּנוּ אֶפֶס זוּלָתוֹ.
כַּכָּתוּב בְּתוֹרָתוֹ, וְיָדַעְתָּ הַיּוֹם
וַהֲשֵׁבֹתָ אֶל לְבָבֶךָ,
כִּי יְיָ הוּא הָאֱלֹהִים
בַּשָּׁמַיִם מִמַּעַל,
וְעַל הָאָרֶץ מִתָּחַת, אֵין עוֹד.

FOR YOU SPREAD OUT THE HEAVENS and established the earth; Your majestic abode is in the heavens above and Your mighty Presence is in the loftiest heights. You are our God and there is none else. In truth You are our Sovereign without compare, as is written in Your Torah: Know then this day and take it to heart that Adonai is surely God in the heavens above and on the earth below. There is none else.

עַל כֵּן נְקַוֶּה לְךָ יְיָ אֱלֹהֵינוּ,
לִרְאוֹת מְהֵרָה בְּתִפְאֶרֶת עֻזֶּךָ,
לְהַעֲבִיר גִּלּוּלִים מִן הָאָרֶץ
וְהָאֱלִילִים כָּרוֹת יִכָּרֵתוּן.
לְתַקֵּן עוֹלָם בְּמַלְכוּת שַׁדַּי,
וְכָל בְּנֵי בָשָׂר יִקְרְאוּ בִשְׁמֶךָ.
לְהַפְנוֹת אֵלֶיךָ כָּל רִשְׁעֵי אָרֶץ.

We therefore hope in You, Adonai our God. May we soon behold the glory of Your might: sweeping away the false gods of the earth that idolatry be utterly destroyed; perfecting the world under the rule of God that all humanity invoke Your name; turning all the wicked of the earth toward You.

אֵין עוֹד *ein od . . . There is none else . . .* The Kabbalah's interpretation is "Adonai is God; there is nothing (!) else," the idea being, "God is all there is!" God and the universe become the same, interwoven in history. *Joel Hoffman*

(שֶׁהוּא) נוֹטֶה שָׁמַיִם *(Shehu) noteh shamayim . . . (For You) spread out the heavens . . .* Isaiah 51:13

וְיָדַעְתָּ הַיּוֹם *V'yadata hayom . . . Know then this day . . .* Deuteronomy 4:39

Aleinu

P In the third paragraph of the *Aleinu,* hope is expressed for a future when all will accept God's rule. Do you think this is meant in the concrete sense, that all people will become Jewish, or is there a different, more metaphoric understanding of accepting God's rule?

עָלֵינוּ

יַכִּירוּ וְיֵדְעוּ כָּל יוֹשְׁבֵי תֵבֵל,
כִּי לְךָ תִּכְרַע כָּל בֶּרֶךְ,
תִּשָּׁבַע כָּל־לָשׁוֹן׃
לְפָנֶיךָ יְיָ אֱלֹהֵינוּ יִכְרְעוּ וְיִפֹּלוּ.
וְלִכְבוֹד שִׁמְךָ יְקָר יִתֵּנוּ.
וִיקַבְּלוּ כֻלָּם אֶת עוֹל מַלְכוּתֶךָ,
וְתִמְלֹךְ עֲלֵיהֶם מְהֵרָה לְעוֹלָם וָעֶד.
כִּי הַמַּלְכוּת שֶׁלְּךָ הִיא,
וּלְעוֹלְמֵי עַד תִּמְלוֹךְ בְּכָבוֹד,
כַּכָּתוּב בְּתוֹרָתֶךָ׃
יְיָ יִמְלֹךְ לְעוֹלָם וָעֶד׃

וְנֶאֱמַר, וְהָיָה יְיָ
לְמֶלֶךְ עַל כָּל הָאָרֶץ.
בַּיּוֹם הַהוּא יִהְיֶה יְיָ אֶחָד
וּשְׁמוֹ אֶחָד.

Let all who dwell on earth acknowledge
that unto You every knee must bend and every tongue swear loyalty.
Before You, Adonai, our God, let them pay homage.
Let them give glory to Your honored Name.
Let all accept the yoke of Your reign,
that You may rule over us soon and forever.
For Sovereignty is Yours
and to all eternity You will reign in glory,
as it is written in Your Torah:
Adonai will reign forever and ever.

Thus it has been said:
Adonai will become Sovereign of all the earth.
On that day Adonai will become One and God's Name will be One.

Kaddish readings begin on page 592. Kaddish is on page 598.

כִּי לְךָ (לִי) תִּכְרַע *Ki l'cha (li) tichra, . . . Unto You (Me) every knee must bend . . .* Isaiah 45:23

יְיָ יִמְלֹךְ *Adonai yimloch . . . Adonai will reign . . .* Exodus 15:18

וְהָיָה יְיָ לְמֶלֶךְ *V'hayah Adonai l'Melech . . . Adonai will become Sovereign . . .* Zechariah 14:9

בַּיּוֹם הַהוּא *Bayom hahu . . . On that day . . .* Zechariah 14:9

Aleinu

קַדִּישׁ יָתוֹם

KADDISH YATOM — MOURNER'S KADDISH

MEDITATIONS BEFORE KADDISH

1.

WHEN I DIE,
Give what's left of me away
To children
And old men that wait to die.
And if you need to cry,
Cry for your brother
Walking the street beside you.
And when you need me,
Put your arms
Around anyone
And give them
What you need to give to me.

I want to leave you something,
Something better . . . Than words
Or sounds.

Look for me
in the people I've known
Or loved,
And if you cannot give me away,
At least let me live on your eyes
And not on your mind.

You can love me most
By letting Hands touch hands,
By letting Bodies touch bodies,
And by letting go
Of children
That need to be free.

Love doesn't die; People do.
So, when all that's left of me
Is love,
Give me away

קַדִּישׁ יָתוֹם *Kaddish Yatom* . . . Mourner's Kaddish . . . The *Kaddish* is a hymn of praise to God and a prayer for the speedy establishment of God's sovereignty on earth, recited at the conclusion of rabbinic study and exposition of Scripture. In its essence it is not a mourner's prayer, and various forms of the *Kaddish* are used to mark the conclusion of each part of the service. The custom of reciting *Kaddish* for a year (or eleven months) after the death of a parent and on the anniversary of that death (*Yahrzeit*) originated in the Rhineland during the Crusades (eleventh century).

Mourner's Kaddish

2.

IN NATURE'S EBB AND FLOW, God's eternal law abides.
When tears dim our vision or grief clouds our understanding,
we often lose sight of God's eternal plan.
Yet we know that growth and decay, life and death,
all reveal a divine purpose.
God who is our support in the struggles of life, is also our hope in death.
We have set God before us and shall not despair.
In God's hands are the souls of all the living and the spirits of all flesh.
Under God's protection we abide, and by God's love are we comforted.
O Life of our life, Soul of our soul, cause Your light to shine into our hearts,
and fill our spirits with abiding trust in You.

3.

THE LIGHT OF LIFE is a finite flame.
Like the Shabbat candles,
life is kindled, it burns, it glows,
it is radiant with warmth and beauty.
But soon it fades, its substance is consumed,
and it is no more.

In light we see;
in light we are seen.
The flames dance and our lives are full.
But as night follows day,
the candle of our life burns down and gutters.
There is an end to the flames.
We see no more
and are no more seen,
yet we do not despair,
for we are more than a memory
slowly fading into the darkness.
With our lives we give life.
Something of us can never die:
we move in the eternal cycle
of darkness and death,
of light and life.

4.

WHY should I wish to see God better than this day?
I see something of God in each hour of the twenty-four,
and each moment then:
In the faces of men and women I see God,
and in my own face in the glass.
I find letters from God dropt in the street,
and every one is sign'd by God's name.
And I leave them where they are,
for I know that whereso'er I go,
others will punctually come forever and ever.

5.

IT IS A FEARFUL THING to love
what death can touch.

A fearful thing to love,
hope, dream: to be —
to be, and oh! to lose.

A thing for fools this, and
a holy thing,
a holy thing to love.

For
your life has lived in me,
your laugh once lifted me,
your word was gift to me.

To remember this brings a painful joy.
'Tis a human thing, love,
a holy thing,
to love
what death has touched.

Mourner's Kaddish

6.

IT IS HARD to sing of oneness when the world is not complete,
when those who once brought wholeness to our life have gone,
and naught but memory can fill the emptiness their passing leaves behind.

But memory can tell us only what we were, in company with those we loved;
it cannot help us find what each of us, alone, must now become.
Yet no one is really alone:
those who live no more, echo still within our thoughts and words,
and what they did is part of what we have become.

We do best homage to our dead when we live our lives more fully,
even in the shadow of our loss.
For each of our lives is worth the life of the whole world;
in each one is the breath of the Ultimate One.
In affirming the One, we affirm the worth of each one
whose life, now ended, brought us closer to the Source of life,
in whose unity no one is alone and every life finds purpose.

7.

יֵשׁ כּוֹכָבִים שֶׁאוֹרָם מַגִּיעַ אַרְצָה
רַק כַּאֲשֶׁר הֵם עַצְמָם אָבְדוּ וְאֵינָם.
יֵשׁ אֲנָשִׁים שֶׁזִּיו זִכְרָם מֵאִיר
כַּאֲשֶׁר הֵם עַצְמָם אֵינָם עוֹד בְּתוֹכֵנוּ.
אוֹרוֹת אֵלֶּה הַמַּבְהִיקִים
בְּחֶשְׁכַּת הַלַּיִל
הֵם הֵם שֶׁמַּרְאִים לָאָדָם אֶת הַדֶּרֶךְ.

THERE ARE STARS up above,
so far away we only see their light
long, long after the star itself is gone.
And so it is with people that we loved —
their memories keep shining ever brightly
though their time with us is done.
But the stars that light up the darkest night,
these are the lights that guide us.
As we live our days, these are the ways we remember.

8.

יִזְכּוֹר We remember

Remember our people who suffered and died so that we could be free and secure;
may their memory be more than a distant shadow.

For their dreams left unfulfilled and lives taken too soon: we remember.

Remember our brothers and sisters whose sacrifice kept the dream of democracy and justice alive; may their courage be our inspiration and strength.

For life cut short and vision unrealized: we remember.

Remember the fallen of our armed services, the victims of terror and tragedy;
may the darkness of their loss not obscure the light of peace. They were in love with our land and in love with life.

For the agony, the tears, the mothers and the fathers,
for the children who were and for the children yet to be: we remember.

Our thoughts turn to those who have departed this earth: our own loved ones, those whom our friends and neighbors have lost, the martyrs of our people whose graves are unmarked, and those of every race and nation whose lives have been a blessing to humanity. As we remember them, we meditate on the meaning of love and loss, of life and death.

Mourner's Kaddish

MOURNER'S KADDISH

קַדִּישׁ יָתוֹם

יִתְגַּדַּל וְיִתְקַדַּשׁ שְׁמֵהּ רַבָּא.
בְּעָלְמָא דִּי בְרָא כִרְעוּתֵהּ,
וְיַמְלִיךְ מַלְכוּתֵהּ,
בְּחַיֵּיכוֹן וּבְיוֹמֵיכוֹן
וּבְחַיֵּי דְכָל בֵּית יִשְׂרָאֵל,
בַּעֲגָלָא וּבִזְמַן קָרִיב. וְאִמְרוּ: אָמֵן.
יְהֵא שְׁמֵהּ רַבָּא מְבָרַךְ
לְעָלַם וּלְעָלְמֵי עָלְמַיָּא.
יִתְבָּרַךְ וְיִשְׁתַּבַּח, וְיִתְפָּאַר
וְיִתְרוֹמַם וְיִתְנַשֵּׂא,
וְיִתְהַדַּר וְיִתְעַלֶּה וְיִתְהַלַּל
שְׁמֵהּ דְּקֻדְשָׁא בְּרִיךְ הוּא,
לְעֵלָּא מִן כָּל בִּרְכָתָא וְשִׁירָתָא,
תֻּשְׁבְּחָתָא וְנֶחֱמָתָא,
דַּאֲמִירָן בְּעָלְמָא. וְאִמְרוּ: אָמֵן.
יְהֵא שְׁלָמָא רַבָּא מִן שְׁמַיָּא,
וְחַיִּים עָלֵינוּ וְעַל כָּל יִשְׂרָאֵל.
וְאִמְרוּ: אָמֵן.
עֹשֶׂה שָׁלוֹם בִּמְרוֹמָיו,
הוּא יַעֲשֶׂה שָׁלוֹם עָלֵינוּ,
וְעַל כָּל יִשְׂרָאֵל. וְאִמְרוּ: אָמֵן.

EXALTED and hallowed be God's great name
in the world which God created, according to plan.
May God's majesty be revealed in the days of our lifetime
and the life of all Israel — speedily, imminently, to which we say Amen.

Blessed be God's great name to all eternity.

Blessed, praised, honored, exalted, extolled, glorified, adored, and lauded
be the name of the Holy Blessed One, beyond all earthly words and songs of blessing,
praise, and comfort. To which we say Amen.

May there be abundant peace from heaven, and life, for us and all Israel.
to which we say Amen.

May the One who creates harmony on high, bring peace to us and to all Israel.
To which we say Amen.

Mourner's Kaddish

I The Mourner's *Kaddish* has been recited by generations of mourners trying to move from grief to acceptance. Listen carefully and hear the voice of those for whom you say *Kaddish,* as they recited this prayer for those who came before, further and further back in time. How does that chorus of mourners provide you with a sense of comfort and strength?

בְּרָכוֹת לַבַּיִת וּלְבֵית הַכְּנֶסֶת

B'RACHOT LABAYIT UL'VEIT HAK'NESSET

BLESSINGS FOR THE HOME AND SYNAGOGUE

בִּרְכוֹת הַמִּשְׁפָּחָה

BIRCHOT HAMISHPACHAH — FAMILY BLESSINGS

IN PRAISE OF A WOMAN

A WOMAN of valor, seek her out;
she is to be valued above rubies.
She opens her hand to those in need
and extends her help to the poor.
Adorned with strength and dignity,
she faces the future cheerfully.
Her speech is wise; the law of kindness is on her lips.
Those who love her rise up with praise and call her blessed:
"Many have done well, but you surpass them all."
Charm is deceptive and beauty short-lived,
but a woman loyal to God has truly earned praise.
Honor her for all of her offerings;
her life proclaims her praise.

IN PRAISE OF A MAN

BLESSED is the man who reveres Adonai,
who greatly delights in God's commandments!
His descendants will be honored in the land
the generation of the upright will be blessed.
His household prospers, and his righteousness endures forever.
Light dawns in the darkness for the upright;
for the one who is gracious, compassionate and just.
He is not afraid of evil tidings;
his mind is firm trusting in Adonai.
His heart is steady, he will not be afraid.
He has distributed freely,
he has given to the poor;
his righteousness endures forever;
his life is exalted in honor.

A woman of valor . . . Proverbs 31
Blessed is the man . . . Psalm 112

Family Blessings

Kiddush, Morning

HaMotzi

Birkat HaMazon

Havdalah

FOR A BOY

יְשִׂימְךָ אֱלֹהִים
כְּאֶפְרַיִם
וְכִמְנַשֶּׁה.

MAY GOD inspire you to live
like Ephraim and Menasseh.

FOR A GIRL

יְשִׂימֵךְ אֱלֹהִים
כְּשָׂרָה, כְּרִבְקָה,
כְּרָחֵל וּכְלֵאָה.

MAY GOD inspire you to live
like Sarah, Rebecca, Rachel and Leah.

FOR BOTH BOYS AND GIRLS

יְבָרֶכְךָ יְיָ וְיִשְׁמְרֶךָ.
יָאֵר יְיָ פָּנָיו אֵלֶיךָ וִיחֻנֶּךָּ.
יִשָּׂא יְיָ פָּנָיו אֵלֶיךָ
וְיָשֵׂם לְךָ שָׁלוֹם.

May God bless you and keep you.
May God's light shine upon you, and may God be gracious to you.
May you feel God's Presence within you always, and may you find peace.

יְבָרֶכְךָ *Y'varech'cha . . . May God bless you . . .* Numbers 6:24–26

בִּרְכוֹת הַמִּשְׁפָּחָה

קִדּוּשׁ, שַׁחֲרִית

הַמּוֹצִיא

בִּרְכַּת הַמָּזוֹן

הַבְדָּלָה

E When we share Shabbat together, we create a special Shabbat family and have an opportunity to share blessings. What blessings would you want to share this Shabbat?

KIDDUSH FOR SHABBAT MORNING

Fill a Kiddush cup with wine or grape juice.
Raise it and recite:

וְשָׁמְרוּ בְנֵי יִשְׂרָאֵל אֶת הַשַּׁבָּת,
לַעֲשׂוֹת אֶת הַשַּׁבָּת לְדֹרֹתָם
בְּרִית עוֹלָם.
בֵּינִי וּבֵין בְּנֵי יִשְׂרָאֵל
אוֹת הִיא לְעוֹלָם,
כִּי שֵׁשֶׁת יָמִים עָשָׂה יְיָ
אֶת הַשָּׁמַיִם וְאֶת הָאָרֶץ,
וּבַיּוֹם הַשְּׁבִיעִי שָׁבַת וַיִּנָּפַשׁ.

THE PEOPLE OF ISRAEL shall keep Shabbat,
observing Shabbat throughout the ages as a covenant for all time.
It shall be a sign for all time between Me and the people of Israel,
for in six days Adonai made heaven and earth,
and on the seventh day God ceased from work and was refreshed.

עַל־כֵּן בֵּרַךְ יְיָ אֶת־יוֹם הַשַּׁבָּת וַיְקַדְּשֵׁהוּ.
Therefore Adonai blessed the day of Shabbat and hallowed it.

בָּרוּךְ אַתָּה, יְיָ אֱלֹהֵינוּ, מֶלֶךְ הָעוֹלָם, בּוֹרֵא פְּרִי הַגָּפֶן.
Praise to You, Adonai our God, Sovereign of the universe, Creator of the fruit of the vine.

וְשָׁמְרוּ בְנֵי יִשְׂרָאֵל *V'shamru v'nei Yisrael . . . The people of Israel shall keep . . .* Exodus 31:16–17

עַל־כֵּן *Al kein . . . Therefore Adonai . . .* Exodus 20:8–11

KIDDUSH FOR FESTIVAL MORNING

ON SHABBAT

וְשָׁמְרוּ בְנֵי יִשְׂרָאֵל אֶת הַשַּׁבָּת,
לַעֲשׂוֹת אֶת הַשַּׁבָּת לְדֹרֹתָם
בְּרִית עוֹלָם.
בֵּינִי וּבֵין בְּנֵי יִשְׂרָאֵל
אוֹת הִיא לְעוֹלָם,
כִּי שֵׁשֶׁת יָמִים עָשָׂה יְיָ
אֶת הַשָּׁמַיִם וְאֶת הָאָרֶץ,
וּבַיּוֹם הַשְּׁבִיעִי שָׁבַת וַיִּנָּפַשׁ.

THE PEOPLE OF ISRAEL shall keep Shabbat, observing Shabbat throughout the ages as a covenant for all time. It shall be a sign for all time between Me and the people of Israel, for in six days Adonai made heaven and earth, and on the seventh day God ceased from work and was refreshed.

עַל־כֵּן בֵּרַךְ יְיָ אֶת־יוֹם הַשַּׁבָּת וַיְקַדְּשֵׁהוּ.
Therefore Adonai blessed the day of Shabbat and hallowed it.

ON ALL FESTIVAL DAYS

אֵלֶּה מוֹעֲדֵי יְיָ,
מִקְרָאֵי קֹדֶשׁ,
אֲשֶׁר תִּקְרְאוּ אֹתָם בְּמוֹעֲדָם.
וַיְדַבֵּר מֹשֶׁה אֶת־מוֹעֲדֵי יְיָ
אֶל־בְּנֵי יִשְׂרָאֵל.

THESE ARE THE APPOINTED SEASONS of Adonai, the sacred days, that you shall proclaim at their appointed times. And Moses declared the appointed seasons of Adonai to the people of Israel.

בָּרוּךְ אַתָּה, יְיָ אֱלֹהֵינוּ,
מֶלֶךְ הָעוֹלָם, בּוֹרֵא פְּרִי הַגָּפֶן.

Praise to You, Adonai our God, Sovereign of the universe, Creator of the fruit of the vine.

וְשָׁמְרוּ *V'shamru . . .* Exodus 31:16–17
עַל־כֵּן *Al kein . . .* Exodus 20:8–11
אֵלֶּה מוֹעֲדֵי *Eileh mo-adei . . .* Leviticus 23:4
וַיְדַבֵּר מֹשֶׁה *Vay'dabeir Moshe . . .* Leviticus 23:44

הַמּוֹצִיא, בִּרְכַּת הַמָּזוֹן

HAMOTZI AND BIRKAT HAMAZON

FOR FOOD

בָּרוּךְ אַתָּה, יְיָ אֱלֹהֵינוּ,
מֶלֶךְ הָעוֹלָם,
הַמּוֹצִיא לֶחֶם מִן הָאָרֶץ.

Our praise to You, Adonai our God, Sovereign of the universe,
who brings forth bread from the earth.

BIRKAT HAMAZON, BLESSING AFTER EATING

ON SHABBAT

שִׁיר הַמַּעֲלוֹת, בְּשׁוּב יְיָ
אֶת־שִׁיבַת צִיּוֹן, הָיִינוּ כְּחֹלְמִים.
אָז יִמָּלֵא שְׂחוֹק,
פִּינוּ וּלְשׁוֹנֵנוּ רִנָּה.
אָז יֹאמְרוּ בַגּוֹיִם,
הִגְדִּיל יְיָ לַעֲשׂוֹת עִם־אֵלֶּה.
הִגְדִּיל יְיָ לַעֲשׂוֹת עִמָּנוּ,
הָיִינוּ שְׂמֵחִים.
שׁוּבָה יְיָ אֶת־שְׁבִיתֵנוּ
כַּאֲפִיקִים בַּנֶּגֶב.
הַזֹּרְעִים בְּדִמְעָה בְּרִנָּה יִקְצֹרוּ.
הָלוֹךְ יֵלֵךְ וּבָכֹה
נֹשֵׂא מֶשֶׁךְ־הַזָּרַע,
בֹּא־יָבוֹא בְרִנָּה נֹשֵׂא אֲלֻמֹּתָיו.

A song of ascents. When Adonai restores the fortunes of Zion, we see it as in a dream, our mouths shall be filled with laughter, our tongues, with songs of joy. Then shall they say among the nations, "Adonai has done great things for them!" Adonai will do great things for us and we shall rejoice. Restore our fortunes, Adonai, like watercourses in the Negev. They who sow in tears shall reap with songs of joy. Those who go forth weeping, carrying the seed-bag, shall come back with songs of joy, carrying their sheaves.

ALL DAYS

Leader

חֲבֵרִים וַחֲבֵרוֹת, נְבָרֵךְ!
Let us praise God.

שִׁיר הַמַּעֲלוֹת *Shir hamaalot . . . A song of ascents . . .* Psalm 126

Group

יְהִי שֵׁם יְיָ מְבֹרָךְ
מֵעַתָּה וְעַד עוֹלָם.

Praised be the name of God, now and forever!

Leader

יְהִי שֵׁם יְיָ מְבֹרָךְ
מֵעַתָּה וְעַד עוֹלָם.
בִּרְשׁוּת הַחֶבְרָה, נְבָרֵךְ אֱלֹהֵינוּ
שֶׁאָכַלְנוּ מִשֶּׁלּוֹ.

Praised be the name of God, now and forever!
Praised be our God, of whose abundance we have eaten.

Group

בָּרוּךְ אֱלֹהֵינוּ שֶׁאָכַלְנוּ מִשֶּׁלּוֹ
וּבְטוּבוֹ חָיִינוּ.

Praised be our God, of whose abundance we have eaten,
and by whose goodness we live.

Leader

בָּרוּךְ אֱלֹהֵינוּ שֶׁאָכַלְנוּ מִשֶּׁלּוֹ
וּבְטוּבוֹ חָיִינוּ.
בָּרוּךְ הוּא וּבָרוּךְ שְׁמוֹ.

Praised be our God, of whose abundance we have eaten,
and by whose goodness we live.
Praised be God and praised be God's name.

Group

בָּרוּךְ אַתָּה, יְיָ אֱלֹהֵינוּ,
מֶלֶךְ הָעוֹלָם, הַזָּן אֶת־הָעוֹלָם
כֻּלּוֹ בְּטוּבוֹ, בְּחֵן בְּחֶסֶד וּבְרַחֲמִים.
הוּא נוֹתֵן לֶחֶם לְכָל־בָּשָׂר,
כִּי לְעוֹלָם חַסְדּוֹ.
וּבְטוּבוֹ הַגָּדוֹל תָּמִיד לֹא חָסַר לָנוּ,
וְאַל יֶחְסַר לָנוּ מָזוֹן לְעוֹלָם וָעֶד.
בַּעֲבוּר שְׁמוֹ הַגָּדוֹל,
כִּי הוּא אֵל זָן וּמְפַרְנֵס לַכֹּל,
וּמֵטִיב לַכֹּל, וּמֵכִין מָזוֹן
לְכָל בְּרִיּוֹתָיו אֲשֶׁר בָּרָא.
בָּרוּךְ אַתָּה, יְיָ, הַזָּן אֶת־הַכֹּל.

Sovereign God of the universe, we praise You: Your goodness sustains the world. You are the God of grace, love, and compassion, the Source of bread for all who live; for Your love is everlasting. In Your great goodness we need never lack for food; You provide food enough for all. We praise You, O God, Source of food for all who live.

כַּכָּתוּב: וְאָכַלְתָּ וְשָׂבָעְתָּ,
וּבֵרַכְתָּ אֶת־יְיָ אֱלֹהֶיךָ
עַל הָאָרֶץ הַטּוֹבָה אֲשֶׁר נָתַן לָךְ.
בָּרוּךְ אַתָּה, יְיָ,
עַל הָאָרֶץ וְעַל הַמָּזוֹן.

As it is written: When you have eaten and are satisfied,
give praise to your God who has given you this good earth.
We praise You, O God, for the earth and for its sustenance.

וּבְנֵה יְרוּשָׁלַיִם עִיר הַקֹּדֶשׁ
בִּמְהֵרָה בְיָמֵינוּ.
בָּרוּךְ אַתָּה, יְיָ,
בּוֹנֶה בְרַחֲמָיו יְרוּשָׁלָיִם. אָמֵן.

Let Jerusalem, the holy city, be renewed in our time.
We praise You, Adonai, in compassion You rebuild Jerusalem. Amen.

הָרַחֲמָן, הוּא יִמְלוֹךְ עָלֵינוּ
לְעוֹלָם וָעֶד.

Merciful One, be our God for ever.

הָרַחֲמָן, הוּא יִתְבָּרַךְ
בַּשָּׁמַיִם וּבָאָרֶץ.

Merciful One, heaven and earth alike are blessed by Your presence.

הָרַחֲמָן, הוּא יִשְׁלַח בְּרָכָה מְרֻבָּה
בַּבַּיִת הַזֶּה,
וְעַל שֻׁלְחָן זֶה שֶׁאָכַלְנוּ עָלָיו.

Merciful One, bless this house and this table at which we have eaten.

ON SHABBAT

הָרַחֲמָן, הוּא יִשְׁלַח לָנוּ
אֶת אֵלִיָּהוּ הַנָּבִיא, זָכוּר לַטּוֹב,
וִיבַשֶּׂר־לָנוּ בְּשׂוֹרוֹת טוֹבוֹת,
יְשׁוּעוֹת וְנֶחָמוֹת.

Merciful One, send us tidings of Elijah, glimpses of good to come, redemption and consolation.

הָרַחֲמָן, הוּא יַנְחִילֵנוּ יוֹם שֶׁכֻּלּוֹ
שַׁבָּת וּמְנוּחָה לְחַיֵּי הָעוֹלָמִים.

Merciful One, help us to see the coming of a time when all is Shabbat.

ON YOM TOV

הָרַחֲמָן, הוּא יַנְחִילֵנוּ
יוֹם שֶׁכֻּלּוֹ טוֹב.

Merciful One, help us to see the coming of a time when all is good.

ALL DAYS

עֹשֶׂה שָׁלוֹם בִּמְרוֹמָיו,
הוּא יַעֲשֶׂה שָׁלוֹם
עָלֵינוּ וְעַל כָּל יִשְׂרָאֵל,
וְעַל כָּל יוֹשְׁבֵי תֵבֵל, וְאִמְרוּ אָמֵן.

May the Source of peace grant peace
to us, to all Israel, and to all the world.

יְיָ עֹז לְעַמּוֹ יִתֵּן.
יְיָ יְבָרֵךְ אֶת־עַמּוֹ בַשָּׁלוֹם.

May Adonai grant strength to our people.
May Adonai bless our people with peace.

עֹשֶׂה שָׁלוֹם *Oseh shalom bimromav . . . May the Source of peace . . .* Job 25:2

יְיָ עֹז *Adonai oz . . . May Adonai grant . . .* Psalm 29:11

HAVDALAH

As Shabbat ends, the Havdalah candle is lit.

הִנֵּה אֵל יְשׁוּעָתִי, אֶבְטַח וְלֹא אֶפְחָד.
כִּי עָזִּי וְזִמְרָת יָהּ יְיָ, וַיְהִי לִי לִישׁוּעָה.
וּשְׁאַבְתֶּם מַיִם בְּשָׂשׂוֹן
מִמַּעַיְנֵי הַיְשׁוּעָה.
לַייָ הַיְשׁוּעָה,
עַל עַמְּךָ בִרְכָתֶךָ, סֶּלָה.
יְיָ צְבָאוֹת עִמָּנוּ,
מִשְׂגָּב לָנוּ אֱלֹהֵי יַעֲקֹב סֶלָה.
יְיָ צְבָאוֹת, אַשְׁרֵי אָדָם בֹּטֵחַ בָּךְ.
יְיָ הוֹשִׁיעָה,
הַמֶּלֶךְ יַעֲנֵנוּ בְיוֹם קָרְאֵנוּ.
לַיְּהוּדִים הָיְתָה
אוֹרָה וְשִׂמְחָה וְשָׂשׂוֹן וִיקָר;
כֵּן תִּהְיֶה לָנוּ.
כּוֹס יְשׁוּעוֹת אֶשָּׂא, וּבְשֵׁם יְיָ אֶקְרָא.

BEHOLD the God who gives me triumph! I am confident, unafraid; for Adonai is my strength and might, and has been my deliverance. Joyfully shall you draw water from the fountains of triumph, deliverance is Adonai's; Your blessing be upon Your people! Selah.

Adonai Tz'vaot is with us; the God of Jacob is our haven. Selah.
Adonai Tz'vaot, happy is the one who trusts in You. O Adonai, grant victory!
May the Sovereign answer us when we call.
The Jews enjoyed light and gladness, happiness and honor. So may it be for us.
I raise the cup of deliverance and invoke the name of Adonai.

הַבְדָּלָה *Havdalah* means *separation*, the separation of Shabbat from the beginning of the new work week.

הִנֵּה אֵל יְשׁוּעָתִי *Hineih El y'shuati . . . Behold the God who gives me triumph! . . .* Shabbat is a foretaste of the world-to-come, a time of perfect peace. These verses call upon God to recall and act on the promise of redemption.

For those who choose: At the last sentence כּוֹס יְשׁוּעוֹת אֶשָּׂא *kos y'shuot esa, I raise the cup of deliverance,* one lifts a cup of wine or grape juice without tasting and proceeds directly to the blessing.

I Think back over the past Shabbat. Remember the moments of strength and the times of rest. What will you draw from them to bring with you through the week to come?

בִּרְכוֹת הַמִּשְׁפָּחָה

קִדּוּשׁ, שַׁחֲרִית

הַמּוֹצִיא

בִּרְכַּת הַמָּזוֹן

הַבְדָּלָה

THE WINE OR GRAPE JUICE

The blessing may be said over wine or grape juice.
Lift the goblet but do not drink until after the Blessing of Separation.

בָּרוּךְ אַתָּה, יְיָ
אֱלֹהֵינוּ, מֶלֶךְ הָעוֹלָם,
בּוֹרֵא פְּרִי הַגָּפֶן.

Praise to You, Adonai our God, Sovereign of the universe, Creator of the fruit of the vine.

THE SPICES

Lift the spice box.

בָּרוּךְ אַתָּה, יְיָ
אֱלֹהֵינוּ, מֶלֶךְ הָעוֹלָם,
בּוֹרֵא מִינֵי בְשָׂמִים.

Praise to You, Adonai our God, Sovereign of the universe, Creator of varied spices.

Circulate the spice box.

THE LIGHT

Raise the Havdalah candle.

בָּרוּךְ אַתָּה, יְיָ
אֱלֹהֵינוּ, מֶלֶךְ הָעוֹלָם,
בּוֹרֵא מְאוֹרֵי הָאֵשׁ.

Praise to You, Adonai our God, Sovereign of the universe, Creator of the lights of fire.

The filled and raised cup is symbolic of the joy felt when expressing gratitude to God. The wine is not sipped until the final blessing.

The spices, coming from the earth, remind us of our duty during our work-week to protect the fragile balance of nature.

I With each ritual involved in *Havdalah,* ready that sense to face the end of Shabbat and the start of something new. With the wine, prepare to taste the new week. With the spices, open your nose to sniff out what is to come. With the candle, open your eyes to see the opportunities. What do you see, as your eyes adjust, after the candle is extinguished in the wine?

THE BLESSING OF SEPARATION

בִּרְכוֹת הַמִּשְׁפָּחָה

קִדּוּשׁ, שַׁחֲרִית

הַמּוֹצִיא

בִּרְכַּת הַמָּזוֹן

הַבְדָּלָה

בָּרוּךְ אַתָּה, יְיָ,
אֱלֹהֵינוּ, מֶלֶךְ הָעוֹלָם,
הַמַּבְדִּיל בֵּין קֹדֶשׁ לְחוֹל,
בֵּין אוֹר לְחְשֶׁךְ,
בֵּין יִשְׂרָאֵל לָעַמִּים,
בֵּין יוֹם הַשְּׁבִיעִי
לְשֵׁשֶׁת יְמֵי הַמַּעֲשֶׂה.
בָּרוּךְ אַתָּה, יְיָ,
הַמַּבְדִּיל בֵּין קֹדֶשׁ לְחוֹל.

PRAISE TO YOU, Adonai our God, Sovereign of the universe:
who distinguishes between the holy and ordinary, between light and dark,
between Israel and the nations, between the seventh day and the six days of work.
Praise to You, Adonai who distinguishes between the holy and ordinary.

Sip the wine or grape juice.

Extinguish the Havdalah candle in the remaining wine or grape juice,
while the following passages are sung or said:

הַמַּבְדִּיל בֵּין קֹדֶשׁ לְחֹל,
חַטֹּאתֵינוּ הוּא יִמְחֹל,
זַרְעֵנוּ וְכַסְפֵּנוּ יַרְבֶּה כַחוֹל,
וְכַכּוֹכָבִים בַּלָּיְלָה.

שָׁבוּעַ טוֹב . . .

MAY THE ONE who distinguishes between the holy and the ordinary, pardon our
sins; multiply our offspring and our possessions as grains of sand and as the stars at night.

A good week, a week of peace, may gladness reign and joy increase.

E Recall a moment of separation you have experienced — a graduation, a trip away from home, moving to a new home. If you were to create a ritual to say farewell to one place and move into another, what would you include?

אֵלִיָּהוּ הַנָּבִיא,
אֵלִיָּהוּ הַתִּשְׁבִּי,
אֵלִיָּהוּ הַגִּלְעָדִי.

בִּמְהֵרָה בְּיָמֵינוּ,
יָבוֹא אֵלֵינוּ,
עִם מָשִׁיחַ בֶּן־דָּוִד.

MAY ELIJAH the prophet,
Elijah the Tishbite,
Elijah of Gilead,
quickly in our day come to us
heralding redemption.

מִרְיָם הַנְּבִיאָה עֹז וְזִמְרָה בְּיָדָהּ.
מִרְיָם תִּרְקֹד אִתָּנוּ
לְהַגְדִּיל זִמְרַת עוֹלָם.
מִרְיָם תִּרְקֹד אִתָּנוּ
לְתַקֵּן אֶת הָעוֹלָם.
בִּמְהֵרָה בְּיָמֵינוּ הִיא תְּבִיאֵנוּ
אֶל מֵי הַיְשׁוּעָה.

MIRIAM the prophet, strength and song are in her hand.
Miriam will dance with us to swell earth's song.
Miriam will dance with us to redeem the world.
Soon, in our day, she will bring us to the waters of redemption.

"Elijah the Tishbite" is a paraphrase of the introduction of the Elijah story in I Kings 17:1.
Marc Brettler

Elijah in Jewish folklore is the champion of the poor and downtrodden, an agent of mercy and hope. He is the bearer of good tidings, the harbinger of the messianic age.

Elijah is not the only figure in Jewish tradition to fulfill a role as God's messenger and harbinger of hope. The sages in the Midrash shaped two biblical women, Serah bat Asher (Genesis 46:17) and Miriam, into such visionary personalities. Each woman gave advice and served as models of redemption for the Jewish people. This prayer which includes Miriam in the Havdalah service is a welcome addition to our liturgy.

שִׁירִים וּזְמִירוֹת

SHIRIM UZ'MIROT

SONGS AND HYMNS

INDEX OF SONGS

This index includes a short selection of songs from the original edition of *Mishkan T'filah*, indicated **in bold.** The other songs, with page numbers from the original edition, are listed here for reference only.

ALPHABETICAL INDEX OF SONGS

Hymns
Shabbat
Havdalah
Songs
Israeli Songs
Meditation and Healing
National Hymns

HYMNS / PIYUTIM

ADON OLAM

אֲדוֹן עוֹלָם אֲשֶׁר מָלַךְ,
בְּטֶרֶם כָּל יְצִיר נִבְרָא.
לְעֵת נַעֲשָׂה בְחֶפְצוֹ כֹּל,
אֲזַי מֶלֶךְ שְׁמוֹ נִקְרָא.

וְאַחֲרֵי כִּכְלוֹת הַכֹּל,
לְבַדּוֹ יִמְלוֹךְ נוֹרָא.
וְהוּא הָיָה, וְהוּא הֹוֶה,
וְהוּא יִהְיֶה, בְּתִפְאָרָה.

וְהוּא אֶחָד וְאֵין שֵׁנִי,
לְהַמְשִׁיל לוֹ לְהַחְבִּירָה.
בְּלִי רֵאשִׁית בְּלִי תַכְלִית,
וְלוֹ הָעֹז וְהַמִּשְׂרָה.

וְהוּא אֵלִי וְחַי גֹּאֲלִי,
וְצוּר חֶבְלִי בְּעֵת צָרָה.
וְהוּא נִסִּי וּמָנוֹס לִי
מְנָת כּוֹסִי בְּיוֹם אֶקְרָא.

בְּיָדוֹ אַפְקִיד רוּחִי,
בְּעֵת אִישַׁן וְאָעִירָה.
וְעִם רוּחִי גְּוִיָּתִי,
יְיָ לִי וְלֹא אִירָא.

You are our Eternal God, who reigned before any being had been created;
when all was done according to Your will, then You were called Ruler.

And after all ceases to be, You alone will rule in majesty.
You have been, are yet, and will be in glory.

And You are One; none other can compare to or consort with You.
You are without beginning, without end. To You belong power and dominion.

And You are my God, my living Redeemer, my Rock in times of trouble and distress.
You are my standard bearer and my refuge, my benefactor when I call on You.

Into Your hands I entrust my spirit, when I sleep and when I wake,
and with my spirit my body also; Adonai is with me and I shall not fear.

EIN K'ELOHEINU

פִּיּוּטִים

שַׁבָּת

הַבְדָּלָה

שִׁירִים

שִׁירֵי אֶרֶץ יִשְׂרָאֵל

מֶדִיטַצְיָה וְהַחְלָמָה

שִׁירִים לָאֻמִּיִּים

אֵין כֵּאלֹהֵינוּ, אֵין כַּאדוֹנֵינוּ,
אֵין כְּמַלְכֵּנוּ, אֵין כְּמוֹשִׁיעֵנוּ.

מִי כֵאלֹהֵינוּ, מִי כַאדוֹנֵינוּ,
מִי כְמַלְכֵּנוּ, מִי כְמוֹשִׁיעֵנוּ?

נוֹדֶה לֵאלֹהֵינוּ, נוֹדֶה לַאדוֹנֵינוּ,
נוֹדֶה לְמַלְכֵּנוּ, נוֹדֶה לְמוֹשִׁיעֵנוּ.

בָּרוּךְ אֱלֹהֵינוּ, בָּרוּךְ אֲדוֹנֵינוּ,
בָּרוּךְ מַלְכֵּנוּ, בָּרוּךְ מוֹשִׁיעֵנוּ.

אַתָּה הוּא אֱלֹהֵינוּ, אַתָּה הוּא אֲדוֹנֵינוּ,
אַתָּה הוּא מַלְכֵּנוּ, אַתָּה הוּא מוֹשִׁיעֵנוּ.

There is none like our God; there is none like our Eternal One;
There is none like our Ruler; there is none like our Redeemer.

Who is like our God? Who is like our Eternal One?
Who is like our Ruler? Who is like our Redeemer?

We will give thanks to our God; we will give thanks to our Eternal One;
We will give thanks to our Ruler; we will give thanks to our Redeemer.

Praised be our God; praised be our Eternal One;
Praised be our Ruler; praised be our Redeemer.

You are our God; You are our Eternal One;
You are our Ruler; You are our Redeemer.

SHALOM ALEICHEM

שָׁלוֹם עֲלֵיכֶם, מַלְאֲכֵי הַשָּׁרֵת,
מַלְאֲכֵי עֶלְיוֹן,
מִמֶּלֶךְ מַלְכֵי הַמְּלָכִים,
הַקָּדוֹשׁ בָּרוּךְ הוּא.

בּוֹאֲכֶם לְשָׁלוֹם, מַלְאֲכֵי הַשָּׁלוֹם,
מַלְאֲכֵי עֶלְיוֹן,
מִמֶּלֶךְ מַלְכֵי הַמְּלָכִים,
הַקָּדוֹשׁ בָּרוּךְ הוּא.

בָּרְכוּנִי לְשָׁלוֹם, מַלְאֲכֵי הַשָּׁלוֹם,
מַלְאֲכֵי עֶלְיוֹן,
מִמֶּלֶךְ מַלְכֵי הַמְּלָכִים,
הַקָּדוֹשׁ בָּרוּךְ הוּא.

צֵאתְכֶם לְשָׁלוֹם, מַלְאֲכֵי הַשָּׁלוֹם,
מַלְאֲכֵי עֶלְיוֹן,
מִמֶּלֶךְ מַלְכֵי הַמְּלָכִים,
הַקָּדוֹשׁ בָּרוּךְ הוּא.

Peace be to you, O ministering angels, messengers of the Most High,
Majesty of majesties, Holy One of Blessing.

Enter in peace, O messengers of peace, angels of the Most High,
Majesty of majesties, Holy One of Blessing.

Bless me with peace, O messengers of peace, angels of the Most High,
Majesty of majesties, Holy One of Blessing.

Depart in peace, O messengers of peace, angels of the Most High,
Majesty of majesties, Holy One of Blessing.

YIGDAL

פִּיּוּטִים

שַׁבָּת

הַבְדָּלָה

שִׁירִים

שִׁירֵי אֶרֶץ יִשְׂרָאֵל

מֶדִיטַצְיָה וְהַחְלָמָה

שִׁירִים לְאֻמִּיִּים

יִגְדַּל אֱלֹהִים חַי וְיִשְׁתַּבַּח,
נִמְצָא, וְאֵין עֵת אֶל מְצִיאוּתוֹ.
אֶחָד, וְאֵין יָחִיד כְּיִחוּדוֹ,
נֶעְלָם, וְגַם אֵין סוֹף לְאַחְדּוּתוֹ.

אֵין לוֹ דְּמוּת הַגּוּף וְאֵינוֹ גוּף,
לֹא נַעֲרוֹךְ אֵלָיו קְדֻשָּׁתוֹ.
קַדְמוֹן לְכָל דָּבָר אֲשֶׁר נִבְרָא,
רִאשׁוֹן, וְאֵין רֵאשִׁית לְרֵאשִׁיתוֹ.

הִנּוֹ אֲדוֹן עוֹלָם, לְכָל נוֹצָר,
יוֹרֶה גְדֻלָּתוֹ וּמַלְכוּתוֹ.
שֶׁפַע נְבוּאָתוֹ נְתָנוֹ, אֶל
אַנְשֵׁי סְגֻלָּתוֹ וְתִפְאַרְתּוֹ.

לֹא קָם בְּיִשְׂרָאֵל כְּמֹשֶׁה עוֹד,
נָבִיא וּמַבִּיט אֶת תְּמוּנָתוֹ.
תּוֹרַת אֱמֶת נָתַן לְעַמּוֹ אֵל,
עַל יַד נְבִיאוֹ נֶאֱמַן בֵּיתוֹ.

לֹא יַחֲלִיף הָאֵל וְלֹא יָמִיר דָּתוֹ,
לְעוֹלָמִים, לְזוּלָתוֹ.
צוֹפֶה וְיוֹדֵעַ סְתָרֵינוּ,
מַבִּיט לְסוֹף דָּבָר בְּקַדְמָתוֹ.

גּוֹמֵל לְאִישׁ חֶסֶד כְּמִפְעָלוֹ,
נוֹתֵן לְרָשָׁע רָע כְּרִשְׁעָתוֹ.
יִשְׁלַח לְקֵץ יָמִין פְּדוּת עוֹלָם,
כָּל חַי וְיֵשׁ יַכִּיר יְשׁוּעָתוֹ.

חַיֵּי עוֹלָם נָטַע בְּתוֹכֵנוּ
בָּרוּךְ עֲדֵי עַד שֵׁם תְּהִלָּתוֹ.

Magnified and praised be the living God; God's existence is Eternal. God's unity is infinite; God is unfathomable, and God's Oneness is unending.

God has neither form nor body; God is incorporeal; God's holiness is beyond compare. God preceded all creation; God is the first and uncreated.

Behold the Eternal, who reveals greatness and sovereignty to every creature. God inspired with the gift of prophecy those chosen to make known God's glory.

Never has there been a prophet like Moses, whose closeness to God is unmatched. A Torah of truth did God give our people, through this prophet, this faithful servant.

God does not change; God's teaching will not be supplanted; God will always be the same. God watches us and knows our secret thoughts; God perceives the end of every matter before it begins.

God deals kindly with those who merit kindness and brings upon the wicked the evil consequences of their deeds. At the end of days, God will send an everlasting redemption; all that lives and breathes will witness God's deliverance.

God has implanted eternal life within us. Blessed is God's glorious Name to all eternity.

ANIM Z'MIROT

אַנְעִים זְמִירוֹת וְשִׁירִים אֶאֱרוֹג
כִּי אֵלֶיךָ נַפְשִׁי תַעֲרוֹג.

נַפְשִׁי חָמְדָה בְּצֵל יָדֶךָ, לָדַעַת,
כָּל־רָז סוֹדֶךָ.

מִדֵּי דַבְּרִי בִּכְבוֹדֶךָ
הוֹמֶה לִבִּי אֶל־דּוֹדֶיךָ
יֶעֱרַב־נָא שִׂיחִי עָלֶיךָ,
כִּי נַפְשִׁי תַעֲרוֹג אֵלֶיךָ.

I seek pleasing melodies and thirst for songs, for my soul thirsts for You.
My soul desires to know what You veil, every mysterious secret.
Whenever I speak of Your glory, my heart yearns for Your love.
Let my prayer be sweet to You, for my soul thirsts for You.

MAH TOVU

פִּיּוּטִים

שַׁבָּת

הַבְדָּלָה

שִׁירִים

שִׁירֵי אֶרֶץ יִשְׂרָאֵל

מֶדִיטַצְיָה וְהַחְלָמָה

שִׁירִים לְאֻמִּיִּים

מַה־טֹּבוּ אֹהָלֶיךָ, יַעֲקֹב,
מִשְׁכְּנֹתֶיךָ, יִשְׂרָאֵל!

וַאֲנִי בְּרֹב חַסְדְּךָ
אָבוֹא בֵיתֶךָ.

אֶשְׁתַּחֲוֶה אֶל־הֵיכַל קָדְשְׁךָ
בְּיִרְאָתֶךָ.

יְיָ, אָהַבְתִּי מְעוֹן בֵּיתֶךָ
וּמְקוֹם מִשְׁכַּן כְּבוֹדֶךָ.

וַאֲנִי אֶשְׁתַּחֲוֶה וְאֶכְרָעָה,
אֶבְרְכָה לִפְנֵי־יְיָ עֹשִׂי.

וַאֲנִי תְפִלָּתִי־לְךָ, יְיָ,
עֵת רָצוֹן.

אֱלֹהִים, בְּרָב־חַסְדֶּךָ,
עֲנֵנִי בֶּאֱמֶת יִשְׁעֶךָ.

How fair are your tents, O Jacob,
your dwellings, O Israel. *(Numbers 24:5)*

I, through Your abundant love, enter Your house.
I bow down in awe at Your holy temple. *(Psalm 5:8)*

Adonai, I love Your temple abode, the dwelling-place of Your glory.
I will humbly bow down low before Adonai, my Maker. *(Psalm 26:8, 95:6)*

As for me, may my prayer come to You, Adonai, at a favorable moment;
O God, in Your abundant faithfulness, answer me with Your sure deliverance.

(Psalm 69:14)

EIN K'ELOHEINU — NON KOMO MUESTRO DIO (SEPHARDIC/LADINO)

אֵין כֵּאלֹהֵינוּ, אֵין כַּאדוֹנֵינוּ, אֵין כְּמַלְכֵּנוּ, אֵין כְּמוֹשִׁיעֵנוּ.

Non komo muestro Dio, non komo muestro Señor,
Non komo muestro Rei, non komo muestro Salvador.

מִי כֵאלֹהֵינוּ, מִי כַאדוֹנֵינוּ, מִי כְמַלְכֵּנוּ, מִי כְמוֹשִׁיעֵנוּ?

Ken komo muestro Dio? Ken komo muestro Señor?
Ken komo muestro Rei? Ken komo muestro Salvador?

נוֹדֶה לֵאלֹהֵינוּ, נוֹדֶה לַאדוֹנֵינוּ, נוֹדֶה לְמַלְכֵּנוּ, נוֹדֶה לְמוֹשִׁיעֵנוּ.

Loaremos a muestro Dio, loaremos a muestro Señor,
Loaremos a muestro Rei, loaremos a muestro Salvador.

בָּרוּךְ אֱלֹהֵינוּ, בָּרוּךְ אֲדוֹנֵינוּ, בָּרוּךְ מַלְכֵּנוּ, בָּרוּךְ מוֹשִׁיעֵנוּ.

Bendicho muestro Dio, bendicho muestro Señor,
Bendicho muestro Rei, bendicho muestro Salvador.

אַתָּה הוּא אֱלֹהֵינוּ, אַתָּה הוּא אֲדוֹנֵינוּ,
אַתָּה הוּא מַלְכֵּנוּ, אַתָּה הוּא מוֹשִׁיעֵנוּ.

Tu sos muestro Dio, Tu sos muestro Señor,
Tu sos muestro Rei, Tu sos muestro Salvador.

EIN ADIR

פִּיּוּטִים

שַׁבָּת

הַבְדָּלָה

שִׁירִים

שִׁירֵי אֶרֶץ יִשְׂרָאֵל

מֶדִיטַצְיָה וְהַחְלָמָה

שִׁירִים לְאֻמִּיִּים

מִפִּי אֵל וּמִפִּי אֵל
יְבֹרַךְ כָּל יִשְׂרָאֵל.

אֵין אַדִּיר כַּיְיָ
וְאֵין בָּרוּךְ כְּבֶן עַמְרָם.
אֵין גְּדוֹלָה כַּתּוֹרָה
וְאֵין דַּרְשָׁנֶיהָ כְּיִשְׂרָאֵל.

אֵין הָדוּר כַּיְיָ
וְאֵין וָתִיק כְּבֶן עַמְרָם.
אֵין זַכָּה כַּתּוֹרָה
וְאֵין חַכָמֶיהָ כְּיִשְׂרָאֵל.

אֵין טָהוֹר כַּיְיָ
וְאֵין יָחִיד כְּבֶן עַמְרָם.
אֵין כַּבִּירָה כַּתּוֹרָה
וְאֵין לַמְדָנֶיהָ כְּיִשְׂרָאֵל.

אֵין פּוֹדֶה כַּיְיָ
וְאֵין צַדִּיק כְּבֶן עַמְרָם.
אֵין קְדוֹשָׁה כַּתּוֹרָה
וְאֵין תּוֹמְכֶיהָ כְּיִשְׂרָאֵל.

From the mouth of God all of Israel is blessed.

There is none as mighty as God, no one as blessed as Amram's son, nothing great as the Torah, and no interpreters like Israel.

There is none as glorified as God, no one as pious as Amram's son, nothing as pure as the Torah, and no scholars like Israel.

There is none as pure as God, no one equal to Amram's son, nothing as mighty as the Torah, and no students like Israel.

There is none who can redeem like God, no one as righteous as Amram's son, nothing as holy as the Torah, and no supporters like Israel.

SHABBAT

YOM ZEH L'YISRAEL

יוֹם זֶה לְיִשְׂרָאֵל אוֹרָה וְשִׂמְחָה,
שַׁבַּת מְנוּחָה.

צִוִּיתָ פִּקּוּדִים בְּמַעֲמַד הַר סִינַי,
שַׁבָּת וּמוֹעֲדִים לִשְׁמֹר בְּכָל שָׁנַי,
לַעֲרֹךְ לְפָנַי מַשְׂאֵת וַאֲרוּחָה,
שַׁבַּת מְנוּחָה.

קִדַּשְׁתָּ, בֵּרַכְתָּ אוֹתוֹ מִכָּל יָמִים,
בְּשֵׁשֶׁת כִּלִּיתָ מְלֶאכֶת עוֹלָמִים.
בּוֹ מָצְאוּ עֲגוּמִים הַשְׁקֵט וּבִטְחָה,
שַׁבַּת מְנוּחָה.

This is Israel's day of light and joy, a Shabbat of rest.

You bade us standing assembled at Mount Sinai,
that all the year through we should keep Shabbat and the festivals:
To set out a full table to honor the Shabbat of rest.

When the work of creating the world was done,
You sanctified and blessed it more than all other days.
On it mourners will find safety and tranquility, a Shabbat of rest.

KI ESHM'RAH SHABBAT

כִּי אֶשְׁמְרָה שַׁבָּת אֵל יִשְׁמְרֵנִי.
אוֹת הִיא לְעוֹלְמֵי עַד בֵּינוֹ וּבֵינִי.

When I keep Shabbat, God keeps me. It is a sign forever between God and me.

MAH YAFEH HAYOM

מַה יָּפֶה הַיּוֹם, שַׁבָּת שָׁלוֹם.

How lovely today is, Shabbat Shalom.

SHABBAT HAMALKAH

הַחַמָּה מֵרֹאשׁ הָאִילָנוֹת נִסְתַּלְּקָה,
בֹּאוּ וְנֵצֵא לִקְרַאת שַׁבָּת הַמַּלְכָּה.
הִנֵּה הִיא יוֹרֶדֶת, הַקְּדוֹשָׁה הַבְּרוּכָה.
וְעִמָּהּ מַלְאָכִים, צְבָא שָׁלוֹם וּמְנוּחָה.
בֹּאִי בֹּאִי הַמַּלְכָּה, בֹּאִי בֹּאִי הַכַּלָּה.
שָׁלוֹם עֲלֵיכֶם, מַלְאֲכֵי הַשָּׁלוֹם.

The sun on the treetops no longer is seen;
come, gather to welcome the Sabbath, our Queen.
Behold her descending, the holy, the blessed,
and with her the angels of peace and of rest.
Draw near, draw near, and here abide;
draw near, draw near, O Sabbath bride.
Peace also to you, you angels of peace.

SHABBAT SHALOM (SHIR HAMAALOT)

— I —

שַׁבָּת שָׁלוֹם.

— II —

שַׁבָּת שָׁלוֹם וּמְבֹרָךְ.

— III —

שִׁיר הַמַּעֲלוֹת בְּשׁוּב יְיָ
אֶת שִׁיבַת צִיּוֹן הָיִינוּ כְּחֹלְמִים.
אָז יִמָּלֵא שְׂחוֹק פִּינוּ וּלְשׁוֹנֵנוּ רִנָּה.
אָז יֹאמְרוּ בַגּוֹיִם
הִגְדִּיל יְיָ לַעֲשׂוֹת עִם אֵלֶּה.

A peaceful Shabbat.

A peaceful and blessed Shabbat.

A song of ascents.
When Adonai returns the exiles of Zion,
we will be like dreamers.
Then our mouths will fill with laughter,
and our tongues with joy.
Then it will be said among the nations,
"Adonai has done great things for them!"

(Psalm 126:1–2)

Y'DID NEFESH

יְדִיד נֶפֶשׁ, אָב הָרַחֲמָן,
מְשׁוֹךְ עַבְדְּךָ אֶל רְצוֹנֶךָ.
יָרוּץ עַבְדְּךָ כְּמוֹ אַיָּל
יִשְׁתַּחֲוֶה אֶל מוּל הֲדָרֶךָ.
יֶעֱרַב לוֹ יְדִידוּתֶךָ
מִנֹּפֶת צוּף וְכָל טָעַם.

הָדוּר נָאֶה, זִיו הָעוֹלָם,
נַפְשִׁי חוֹלַת אַהֲבָתֶךָ:
אָנָּא אֵל, נָא רְפָא נָא לָהּ,
בְּהַרְאוֹת לָהּ נֹעַם זִיוֶךָ.
אָז תִּתְחַזֵּק וְתִתְרַפֵּא,
וְהָיְתָה לָהּ שִׂמְחַת עוֹלָם.

וָתִיק, יֶהֱמוּ נָא רַחֲמֶיךָ,
וְחוּסָה נָּא עַל בֵּן אֲהוּבֶךָ.
כִּי זֶה כַּמָּה נִכְסֹף נִכְסַפְתִּי
לִרְאוֹת בְּתִפְאֶרֶת עֻזֶּךָ.
אָנָּא, אֵלִי, חֶמְדַּת לִבִּי,
חוּסָה נָּא וְאַל נָא תִּתְעַלָּם.

הִגָּלֶה נָא, וּפְרוֹשׂ, חֲבִיבִי עָלַי,
אֶת סֻכַּת שְׁלוֹמֶךָ.
תָּאִיר אֶרֶץ מִכְּבוֹדֶךָ;
נָגִילָה וְנִשְׂמְחָה בָּךְ.
מַהֵר, אָהוּב, כִּי בָא מוֹעֵד,
וְחָנֵּנוּ כִּימֵי עוֹלָם.

Heart's delight, Source of mercy,
draw Your servant into Your arms.
I leap like a deer
to stand in awe before You.
Your love is sweeter to me
than the taste of honey.

World's light, shining glory,
my heart is faint for love of You.
Heal it, O God; help my heart;
show me Your radiant splendor.
Let me return to strength
and have joy for ever.

Have compassion, O Faithful One,
pity for Your loved child.
How long have I hoped
to see Your glorious might.
O God, my heart's desire,
have pity and hold back no more.

Show Yourself, Beloved, and cover me
with the shelter of Your peace.
Light up the world with Your Presence
that we may exult and rejoice in You.
Hurry, Loved One, the appointed time
has come. Show us grace as long ago.

TZUR MISHELO

צוּר, מִשֶּׁלּוֹ אָכַֽלְנוּ, בָּרְכוּ אֱמוּנַי;
שָׂבַֽעְנוּ וְהוֹתַֽרְנוּ כִּדְבַר יְיָ.

הַזָּן אֶת עוֹלָמוֹ רוֹעֵֽנוּ, אָבִֽינוּ;
אָכַֽלְנוּ אֶת לַחְמוֹ וְיֵינוֹ שָׁתִֽינוּ.

עַל כֵּן, נוֹדֶה לִשְׁמוֹ וּנְהַלְלוֹ בְּפִֽינוּ
אָמַֽרְנוּ וְעָנִֽינוּ: אֵין קָדוֹשׁ כַּייָ.

My faithful ones, bless the Rock, from whose food you have eaten.
We ate and have some left, according to God's word.
You nourish Your world, Protective Shepherd, so that we may eat Your bread and drink Your wine. We give You thanks and praise, responding to Your goodness with: "There is no Holy One like Adonai."

YAH RIBON

יָהּ רִבּוֹן עָלַם וְעָלְמַיָּא,
אַנְתְּ הוּא מַלְכָּא, מֶֽלֶךְ מַלְכַיָּא.
עוֹבַד גְּבוּרְתֵּךְ, וְתִמְהַיָּא,
שְׁפַר קֳדָמַי, לְהַחֲוָיָה.

שְׁבָחִין אֲסַדֵּר צַפְרָא וְרַמְשָׁא,
לָךְ אֱלָהָא קַדִּישָׁא דִּי בְרָא כָּל נַפְשָׁא,
עִירִין קַדִּישִׁין וּבְנֵי אֱנָשָׁא,
חֵיוַת בָּרָא וְעוֹפֵי שְׁמַיָּא.

God of this and all worlds, You are supreme, the Sovereign God.
Your mighty, wondrous work moves my heart to praise You.

I will offer praises morning and night, for You, holy God, who has created all life, holy beings and humans, animals on land and birds in the sky.

MIZMOR SHIR L'YOM HASHABBAT

פִּיּוּטִים

שַׁבָּת

הַבְדָּלָה

שִׁירִים

שִׁירֵי אֶרֶץ יִשְׂרָאֵל

מֶדִיטַצְיָה וְהַחְלָמָה

שִׁירִים לְאֻמִּיִּים

מִזְמוֹר שִׁיר לְיוֹם הַשַּׁבָּת.
טוֹב לְהֹדוֹת לַיָי
וּלְזַמֵּר לְשִׁמְךָ עֶלְיוֹן.

לְהַגִּיד בַּבֹּקֶר חַסְדֶּךָ,
וֶאֱמוּנָתְךָ בַּלֵּילוֹת.
עֲלֵי־עָשׂוֹר וַעֲלֵי־נָבֶל, עֲלֵי הִגָּיוֹן בְּכִנּוֹר.

כִּי שִׂמַּחְתַּנִי יְיָ בְּפָעֳלֶךָ
בְּמַעֲשֵׂי יָדֶיךָ אֲרַנֵּן׃
מַה־גָּדְלוּ מַעֲשֶׂיךָ, יְיָ,
מְאֹד עָמְקוּ מַחְשְׁבֹתֶיךָ.

A Psalm, a song for Shabbat.
It is good to praise Adonai,
to sing hymns to Your name, O Most High.

To proclaim Your steadfast love at daybreak,
Your faithfulness each night
with a ten-stringed harp,
with voice and lyre together.

You have gladdened me by Your deeds, Adonai.
I shout for joy at Your handiwork.
How great are Your works, Adonai,
how very profound Your designs! *(Psalm 92:1–6)*

HAVDALAH

LAY'HUDIM HAY'TAH ORAH

לַיְּהוּדִים הָיְתָה אוֹרָה וְשִׂמְחָה
וְשָׂשֹׂן וִיקָר.

The Jews had light, joy, gladness, and honor. *(Esther 8:16)*

Hymns
Shabbat
Havdalah
Songs
Israeli Songs
Meditation and Healing
National Hymns

LANEIR V'LIV'SAMIM

לַנֵּר וְלִבְשָׂמִים נַפְשִׁי מְיַחֲלָה
אִם תִּתְּנוּ לִי כּוֹס יַיִן לְהַבְדָּלָה.

עֵינַי אֲנִי אֶשָּׂא, אֶל־עַל בְּלֵב כּוֹסֵף
מַמְצִיא צְרָכַי לִי, בַּיּוֹם וּבַלָּיְלָה.

My soul awaits the candle and the spices,
if you should but give me a cup of wine for Havdalah.

I will lift up my eyes with a full heart to God,
who fulfills my every need day and night.

ELIYAHU HANAVI

אֵלִיָּהוּ הַנָּבִיא,
אֵלִיָּהוּ הַתִּשְׁבִּי,
אֵלִיָּהוּ הַגִּלְעָדִי.

בִּמְהֵרָה בְיָמֵינוּ,
יָבוֹא אֵלֵינוּ,
עִם מָשִׁיחַ בֶּן־דָּוִד.

May Elijah the prophet, Elijah of Tishbi, Elijah of Gilead,
quickly in our day come to us, heralding redemption.

SHAVUA TOV, MAY YOU HAVE A GOOD WEEK

The twisted candle brightens our hearts as together we watch the Sabbath depart.
We smell the spices, taste the wine, as the stars in the sky begin to shine.

> *Shavua tov,* may you have a good week,
> may you find the happiness you seek.
> *Shavua tov,* may your week be fine,
> may it be as sweet as the Sabbath wine.

We say goodbye to a special friend, another Shabbat has come to an end.
"Shavua tov" are the words we speak to say, "May you have a happy week."

SONGS

ADONAI OZ

יְיָ עֹז לְעַמּוֹ יִתֵּן,
יְיָ יְבָרֵךְ אֶת־עַמּוֹ בַשָּׁלוֹם.

May Adonai grant strength to our people;
may Adonai bless our people with peace. *(Psalm 29:11)*

AL SH'LOSHAH D'VARIM

עַל שְׁלשָׁה דְּבָרִים הָעוֹלָם עוֹמֵד:
עַל הַתּוֹרָה, וְעַל הָעֲבוֹדָה
וְעַל גְּמִילוּת חֲסָדִים.

The world is sustained by three things: Torah, worship and loving deeds.
(Pirkei Avot 1:2)

AM YISRAEL CHAI

עַם יִשְׂרָאֵל חַי! עוֹד אָבִינוּ חַי!

The people of Israel lives! Our God yet lives!

BLESSING FOR SOCIAL JUSTICE

בָּרוּךְ אַתָּה יְיָ אֱלֹהֵינוּ מֶלֶךְ הָעוֹלָם,
אֲשֶׁר קִדְּשָׁנוּ בְּמִצְוֹתָיו,
וְצִוָּנוּ לִרְדּוֹף צֶדֶק.

Blessed is the Eternal our God, Ruler of the universe, who hallows us with mitzvot, commanding us to pursue justice.

EILEH CHAMDAH LIBI

אֵלֶּה חָמְדָה לִבִּי
חוּסָה נָא וְאַל נָא תִּתְעַלָּם.

God is my heart's desire. Appear! Do not hide.

ESA EINAI

אֶשָּׂא עֵינַי אֶל־הֶהָרִים, מֵאַיִן יָבֹא עֶזְרִי?
עֶזְרִי מֵעִם יְיָ, עֹשֵׂה שָׁמַיִם וָאָרֶץ.

I lift my eyes to the mountains; from where does my help come?
My help comes from God, maker of heaven and earth. *(Psalm 121:1–2)*

GESHER TZAR M'OD

כָּל הָעוֹלָם כֻּלּוֹ גֶּשֶׁר צַר מְאֹד
וְהָעִקָּר לֹא לְפַחֵד כְּלָל.

The entire world is but a narrow bridge; the most important thing is not to be afraid.

HAL'LI

הַלְלִי נַפְשִׁי אֶת־יְיָ.
אֲהַלְלָה יְיָ בְּחַיָּי,
אֲזַמְּרָה לֵאלֹהַי בְּעוֹדִי.

Praise Adonai, O my soul!
I will praise Adonai all my life, sing hymns to my God while I exist.
(Psalm 146:1–2)

HAL'LU

הַלְלוּ . . .
כֹּל הַנְּשָׁמָה תְּהַלֵּל יָהּ, הַלְלוּ, הַלְלוּ־יָהּ.

Let all that breathes praise Adonai. Hallelujah. *(Psalm 150:6)*

HAVAH NASHIRAH

הָבָה נָשִׁירָה שִׁיר הַלְלוּ־יָהּ.

Let us sing a song of praise. Hallelujah.

פִּיּוּטִים

שַׁבָּת

הַבְדָּלָה

שִׁירִים

שִׁירֵי אֶרֶץ יִשְׂרָאֵל

מֶדִיטַצְיָה וְהַחְלָמָה

שִׁירִים לְאֻמִּיִּם

HEIVEINU SHALOM ALEICHEM

הֵבֵאנוּ שָׁלוֹם עֲלֵיכֶם,
עֲלֵיכֶם הֵבֵאנוּ שָׁלוֹם.

Let us bring you peace.

HINEIH MAH TOV

הִנֵּה מַה־טּוֹב וּמַה־נָּעִים
שֶׁבֶת אַחִים גַּם־יַחַד.

How good and how pleasant it is that brothers/sisters dwell together.

(Psalm 133:1)

IM EIN ANI LI MI LI?

אִם אֵין אֲנִי לִי, מִי לִי?
וּכְשֶׁאֲנִי לְעַצְמִי, מָה אֲנִי?
וְאִם לֹא עַכְשָׁיו, אֵימָתַי?

If I am not for myself, who will be for me?
And if I am for myself, what am I?
And if not now, when? *(Pirkei Avot 1:14)*

IM TIRTZU

אִם תִּרְצוּ, אֵין זוֹ אַגָּדָה
לִהְיוֹת עַם חָפְשִׁי בְּאַרְצֵנוּ,
בְּאֶרֶץ צִיּוֹן, וִירוּשָׁלָיִם.

If you will it, it is no dream.
To be a free people in our land of Zion and Jerusalem.

IVDU ET ADONAI B'SIMCHAH

עִבְדוּ אֶת יְיָ בְּשִׂמְחָה,
בֹּאוּ לְפָנָיו בִּרְנָנָה.

Worship God with gladness! Come into God's presence with singing.

(Psalm 100:2)

L'CHI LACH

לְכִי לָךְ, לֶךְ־לְךָ.
לְשִׂמְחַת חַיִּים, לְכִי לָךְ.

L'chi lach to a land that I will show you,
Lech l'cha to a place you do not know,
L'chi lach, on your journey I will bless you,
And you shall be a blessing, you shall be a blessing,
You shall be a blessing, *l'chi lach.*

L'chi lach and I shall make your name great,
Lech l'cha and all shall praise your name,
L'chi lach, to the place that I will show you,
L'simchat chayim (3x), *l'chi lach.*

LO ALECHA

לֹא עָלֶיךָ הַמְּלָאכָה לִגְמוֹר
וְלֹא אַתָּה בֶן חוֹרִין לְהִבָּטֵל מִמֶּנָּה.

It is not your duty to complete the work.
Neither are you free to desist from it.

(Pirkei Avot 2:16)

LO YAREI-U

לֹא־יָרֵעוּ וְלֹא־יַשְׁחִיתוּ
וְכִתְּתוּ חַרְבוֹתָם לְאִתִּים
וַחֲנִיתוֹתֵיהֶם לְמַזְמֵרוֹת.

They shall not hurt or destroy.
They shall beat their swords into plowshares,
and their spears into pruning hooks.

(Isaiah 11:9, 2:4)

LO YISA GOI

לֹא־יִשָּׂא גוֹי אֶל־גּוֹי חֶרֶב,
וְלֹא־יִלְמְדוּ עוֹד מִלְחָמָה.

Nation shall not take up sword against nation; they shall never again study war.

(Isaiah 2:4)

MAH GADLU

פִּיּוּטִים

שַׁבָּת

הַבְדָּלָה

שִׁירִים

שִׁירֵי אֶרֶץ יִשְׂרָאֵל

מֶדִיטַצְיָה וְהַחְלָמָה

שִׁירִים לְאֻמִּיִּים

מַה־גָּדְלוּ מַעֲשֶׂיךָ יָהּ,
מְאֹד עָמְקוּ מַחְשְׁבֹתֶיךָ.

How great are Your works, Adonai,
how very profound Your designs!   *(Psalm 92:6)*

MIRIAM'S SONG

And the women dancing with their timbrels
Followed Miriam as she sang her song,
Sing a song to the One whom we've exalted,
Miriam and the woman danced and danced the whole night long.

And Miriam was a weaver of unique variety,
The tapestry she wove was one which sang our history.
With every thread and every strand she crafted her delight.
A woman touched with spirit, she dances toward the light.

As Miriam stood upon the shores and gazed across the seas
The wonder of this miracle she soon came to believe.
Whoever thought the sea would part, with an outstretched hand
And we would pass to freedom, and march to the promised land?

And Miriam the prophet took her timbrel in her hand
And all the women followed her, just as she had planned.
And Miriam raised her voice in song, she sang with praise and might:
We've just lived through a miracle, we're going to dance tonight!

MITZVAH GORERET MITZVAH

מִצְוָה גּוֹרֶרֶת מִצְוָה,
עֲבֵרָה גּוֹרֶרֶת עֲבֵרָה.
לִהְיוֹת צַדִּיק זֶה טוֹב מְאֹד.

One mitzvah leads to another.
One sin leads to another.   *(Pirkei Avot 4:2)*
To be righteous is very good.

MODEH / MODAH ANI

מוֹדֶה/מוֹדָה אֲנִי לְפָנֶיךָ,
מֶלֶךְ חַי וְקַיָּם,
שֶׁהֶחֱזַרְתָּ בִּי נִשְׁמָתִי בְּחֶמְלָה,
רַבָּה אֱמוּנָתֶךָ.

I offer thanks to You, ever-living Sovereign,
that You have restored my soul to me in mercy; how great is Your trust.

OD YISHAMA

עוֹד יִשָּׁמַע בְּעָרֵי יְהוּדָה
וּבְחֻצוֹת יְרוּשָׁלַיִם
קוֹל שָׂשׂוֹן וְקוֹל שִׂמְחָה,
קוֹל חָתָן וְקוֹל כַּלָּה.

There shall yet be heard in the cities of Judah and the outskirts of Jerusalem the sounds of gladness and joy, the voice of bridegroom and bride.

(abridgement of Jeremiah 33:10-11)

OR ZARUA

אוֹר זָרֻעַ לַצַּדִּיק וּלְיִשְׁרֵי־לֵב שִׂמְחָה.

Light is sown for the righteous, joy for the upright. *(Psalm 97:11)*

OSEH SHALOM

עֹשֶׂה שָׁלוֹם בִּמְרוֹמָיו,
הוּא יַעֲשֶׂה שָׁלוֹם עָלֵינוּ
וְעַל כָּל יִשְׂרָאֵל, וְאִמְרוּ: אָמֵן.

May the One who causes peace to reign in the high heavens let peace descend on us and on all Israel, and let us say: Amen.

OZI V'ZIMRAT YAH

עָזִּי וְזִמְרָת יָהּ, וַיְהִי־לִי לִישׁוּעָה.

Adonai is my strength and might; God will be my salvation. *(Exodus 15:2)*

PITCHU LI

פִּיּוּטִים

שַׁבָּת

הַבְדָּלָה

שִׁירִים

שִׁירֵי אֶרֶץ יִשְׂרָאֵל

מֶדִיטַצְיָה וְהַחְכָּמָה

שִׁירִים לְאֻמִּיִּים

פִּתְחוּ־לִי שַׁעֲרֵי־צֶדֶק
אָבֹא־בָם אוֹדֶה יָהּ.

Open the gates of righteousness for me that I may enter them and praise Adonai.

(Psalm 118:19)

SHEHECHEYANU

בָּרוּךְ אַתָּה, יְיָ אֱלֹהֵינוּ, מֶלֶךְ הָעוֹלָם,
שֶׁהֶחֱיָנוּ וְקִיְּמָנוּ וְהִגִּיעָנוּ
לַזְּמַן הַזֶּה.

Praise to You, Adonai our God, Sovereign of the universe,
for giving us life, sustaining us, and enabling us to reach this season.

SHIR CHADASH

שִׁירוּ לַיְיָ כָּל הָאָרֶץ
שִׁירוּ לַיְיָ שִׁיר חָדָשׁ.

Sing unto God, all the earth, a new song.
I will sing unto God a new song.
Sing unto God and we'll all sing along,
all the earth, a new song, unto God.

(based on Psalm 96:1)

SIMAN TOV UMAZAL TOV

סִימָן טוֹב וּמַזָּל טוֹב.
וּמַזָּל טוֹב וְסִימָן טוֹב.
יְהֵי לָנוּ.
יְהֵי לָנוּ, יְהֵי לָנוּ, וּלְכָל יִשְׂרָאֵל.

It is a good and lucky sign for us and all Israel!

T'FILAT HADERECH

May we be blessed as we go on our way.
May we be guided in peace.
May we be blessed with health and joy.
May this be our blessing, Amen.

May we be sheltered by the wings of peace.
May we be kept in safety and in love.
May grace and compassion find their way to every soul.
May this be our blessing, Amen.

THIS IS VERY GOOD

When God made the world and made it full of light,
The sun to shine by day, the moon and stars by night,
God made it full of life, lilies, oaks, and trout;
Tigers and bears; sparrows, hawks and apes.

And God took clay from Earth's four corners
To give it the breath of life, and God said:

> This is very good, this is very good,
> man, woman, and child, all are good.

Man, woman, and child resemble God.
Like God we love, like God we think, like God we care.

V'EIZEHU

וְאֵיזֶהוּ חָכָם, הַלּוֹמֵד מִכָּל אָדָם.
וְאֵיזֶהוּ גִּבּוֹר, הַכּוֹבֵשׁ אֶת יִצְרוֹ.
וְאֵיזֶהוּ עָשִׁיר, הַשָּׂמֵחַ בְּחֶלְקוֹ.

Who are wise? Those who learn from everyone.
Who are mighty? Those who control their urges.
Who are rich? Those who are happy with what they have.

(based on Pirkei Avot 4:1)

פִּיּוּטִים

שַׁבָּת

הַבְדָּלָה

שִׁירִים

שִׁירֵי אֶרֶץ יִשְׂרָאֵל

מֶדִיטַצְיָה וְהַחְלָמָה

שִׁירִים לְאֻמִּיִּים

V'HA-EIR EINEINU

וְהָאֵר עֵינֵינוּ בְּתוֹרָתֶךָ,

וְדַבֵּק לִבֵּנוּ בְּמִצְוֹתֶיךָ.

וְיַחֵד לְבָבֵנוּ,

לְאַהֲבָה וּלְיִרְאָה אֶת־שְׁמֶךָ,

וְלֹא נֵבוֹשׁ, וְלֹא נִכָּלֵם,

וְלֹא נִכָּשֵׁל לְעוֹלָם וָעֶד.

Enlighten our eyes with Your Torah. Cause our hearts to cling to Your commandments. Unite our hearts to love and revere Your name, so that we may never be put to shame.

V'NOMAR L'FANAV

וְנֹאמַר לְפָנָיו שִׁירָה חֲדָשָׁה. הַלְלוּ־יָהּ.

Sing before God a new song, Hallelujah.

V'TAHEIR LIBEINU

וְטַהֵר לִבֵּנוּ לְעָבְדְּךָ בֶּאֱמֶת.

Purify our hearts to serve You in truth. *(based on Proverbs 20:9)*

V'YASHVU ISH

וְיָשְׁבוּ אִישׁ תַּחַת גַּפְנוֹ

וְתַחַת תְּאֵנָתוֹ וְאֵין מַחֲרִיד.

And all shall sit under their vines
and fig trees, and none shall make them afraid. *(Micah 4:4)*

YISM'CHU HASHAMAYIM

יִשְׂמְחוּ הַשָּׁמַיִם וְתָגֵל הָאָרֶץ.

יִרְעַם הַיָּם וּמְלֹאוֹ.

Let the heavens rejoice and the earth exult.
Let the sea roar and everything within it. *(Psalm 96:11)*

ISRAELI SONGS

AL KOL EILEH

עַל הַדְּבַשׁ וְעַל הָעֹקֶץ, עַל הַמַּר וְהַמָּתוֹק
עַל בִּתֵּנוּ הַתִּינֹוקֶת, שְׁמֹר אֵלִי הַטּוֹב.
עַל הָאֵשׁ הַמְבֹעֶרֶת, עַל הַמַּיִם הַזַּכִּים.
עַל הָאִישׁ הַשָּׁב הַבַּיְתָה מִן הַמֶּרְחַקִּים.

עַל כָּל אֵלֶּה, עַל כָּל אֵלֶּה,
שְׁמֹר נָא לִי אֵלִי הַטּוֹב.
עַל הַדְּבַשׁ וְעַל הָעֹקֶץ
עַל הַמַּר וְהַמָּתוֹק.
אַל נָא תַּעֲקוֹר נָטוּעַ,
אַל תִּשְׁכַּח אֶת הַתִּקְוָה.
הֲשִׁיבֵנִי וְאָשׁוּבָה
אֶל הָאָרֶץ הַטּוֹבָה.

שְׁמֹר אֵלִי עַל זֶה הַבַּיִת,
עַל הַגַּן, עַל הַחוֹמָה
מִיָּגוֹן מִפַּחַד פֶּתַע וּמִמִּלְחָמָה.
שְׁמֹר עַל הַמְּעַט שֶׁיֵּשׁ לִי,
עַל הָאוֹר וְעַל הַטַּף
עַל הַפְּרִי שֶׁלֹּא הִבְשִׁיל עוֹד וְשֶׁנֶּאֱסַף.

מְרַשְׁרֵשׁ אִילָן בָּרוּחַ,
מֵרָחוֹק נוֹשֵׁר כּוֹכָב
מִשְׁאֲלוֹת לִבִּי בַּחֹשֶׁךְ נִרְשָׁמוֹת עַכְשָׁיו.
אָנָּא שְׁמֹר לִי עַל כָּל אֵלֶּה
וְעַל אֲהוּבֵי נַפְשִׁי
עַל הַשֶּׁקֶט עַל הַבֶּכִי וְעַל זֶה הַשִּׁיר.

My Good God, keep these safe: the honey and sting, the bitter and the sweet, and our baby daughter; the burning flame, the pure water, and the man returning home from afar.

Keep all of these safe my Good God: the honey and the sting, the bitter and the sweet. Do not uproot what has been planted; do not forget the hope. Return me, and I will return to the good land.

[Cont.]

פִּיּוּטִים

שַׁבָּת

הַבְדָּלָה

שִׁירִים

שִׁירֵי אֶרֶץ יִשְׂרָאֵל

מֶדִיטַצְיָה וְהַחְלָמָה

שִׁירִים לְאֻמִּיִּים

My God, keep this house, this garden and this wall from sorrow, from unexpected fear and from war. Keep safe what little I have, the light and the children, and the ripened fruit that has yet to be picked.

A tree rustles in the wind. A star cascades in the distance. And now my heart's desires are recorded in the darkness. Please keep all of these safe for me. Keep safe the ones I love, the quiet, the crying, and this very song.

ANI V'ATAH

אֲנִי וְאַתָּה נְשַׁנֶּה אֶת הָעוֹלָם,
אֲנִי וְאַתָּה, אָז יָבוֹאוּ כְּבָר כֻּלָּם.

אָמְרוּ אֶת זֶה קֹדֶם לְפָנַי,
(זֶה) לֹא מְשַׁנֶּה
אֲנִי וְאַתָּה נְשַׁנֶּה אֶת הָעוֹלָם.

אֲנִי וְאַתָּה נְנַסֶּה מֵהַתְחָלָה;
יִהְיֶה לָנוּ רַע, אֵין דָּבָר, זֶה לֹא נוֹרָא.

You and I will change the world, you and I. Then all will join us. Though it's been said before it doesn't matter. You and I will change the world. You and I will start from the beginning. It may be difficult, but it doesn't matter.

BASHANAH HABAAH

בַּשָּׁנָה הַבָּאָה נֵשֵׁב עַל הַמִּרְפֶּסֶת
וְנִסְפֹּר צִפֳּרִים נוֹדְדוֹת.
יְלָדִים בַּחֻפְשָׁה יְשַׂחֲקוּ תּוֹפֶסֶת
בֵּין הַבַּיִת לְבֵין הַשָּׂדוֹת.

עוֹד תִּרְאֶה, עוֹד תִּרְאֶה
כַּמָּה טוֹב יִהְיֶה
בַּשָּׁנָה בַּשָּׁנָה הַבָּאָה.

עֲנָבִים אֲדֻמִּים יַבְשִׁילוּ עַד הָעֶרֶב
וְיֻגְּשׁוּ צוֹנְנִים לַשֻּׁלְחָן.
וְרוּחוֹת רְדוּמִים יִשְׂאוּ עַל אֵם הַדֶּרֶךְ
עִתּוֹנִים יְשָׁנִים כֶּעָנָן.

בַּשָּׁנָה הַבָּאָה נִפְרֹשׂ כַּפּוֹת יָדַיִם
מוּל הָאוֹר הַנִּגָּר הַלָּבָן.
אֲנָפָה לְבָנָה תִּפְרֹשׂ בָּאוֹר כְּנָפַיִם
וְהַשֶּׁמֶשׁ תִּזְרַח בְּתוֹכָן.

Next year we will sit on the porch and count migrating birds.
Children on vacation will play catch between the house and the fields.

You will yet see how good it will be next year.

Red grapes will ripen till the evening and will be served chilled to the table.
And lazy winds will carry to the crossroads old newpapers like a cloud.

Next year we will spread our own hands toward the radiant light.
A white heron will spread her wings in the light as the sun shines through them.

DODI LI

דּוֹדִי לִי וַאֲנִי לוֹ, הָרֹעֶה בַּשּׁוֹשַׁנִּים.
מִי זֹאת עֹלָה מִן־הַמִּדְבָּר,
מְקֻטֶּרֶת מוֹר וּלְבוֹנָה?
לִבַּבְתִּינִי, אֲחֹתִי כַלָּה.
עוּרִי צָפוֹן וּבוֹאִי תֵימָן.

My beloved is mine and I am my beloved's, a shepherd among the lilies.
Who is this that comes up from the desert, fragrant with myrrh and frankincense?
You have captured my heart, my sister, my bride.
Awake, O north wind. Come, O south wind!

(Song of Songs 2:16; 3:6; 4:9, 16)

ELI ELI (HALICHAH L'KEISARIAH)

אֵלִי אֵלִי, שֶׁלֹּא יִגָּמֵר לְעוֹלָם:
הַחוֹל וְהַיָּם, רִשְׁרוּשׁ שֶׁל הַמַּיִם,
בְּרַק הַשָּׁמַיִם, תְּפִלַּת הָאָדָם.

O God, my God, I pray that these things never end
The sand and the sea, the rush of the waters,
The crash of the heavens, the prayer of the heart.

ERETZ ZAVAT CHALAV

אֶרֶץ זָבַת חָלָב וּדְבַשׁ.

A land flowing with milk and honey. *(Exodus 3:8)*

EREV SHEL SHOSHANIM

עֶרֶב שֶׁל שׁוֹשַׁנִּים נֵצֵא נָא אֶל הַבֻּסְתָּן.
מוֹר, בְּשָׂמִים וּלְבוֹנָה לְרַגְלֵךְ מִפְתָּן.

לַיְלָה יוֹרֵד לְאַט וְרוּחַ שׁוֹשָׁן נוֹשְׁבָה
הָבָה אֶלְחַשׁ לָךְ שִׁיר בַּלָּאט,
זֶמֶר שֶׁל אַהֲבָה.

שַׁחַר הוֹמָה יוֹנָה רֹאשֵׁךְ מָלֵא טְלָלִים
פִּיךְ אֶל הַבֹּקֶר שׁוֹשַׁנָּה אֶקְטְפֶנּוּ לִי.

It is an evening of roses. Let us go out to the grove.
Myrrh, spices, and frankincense are a carpet for you to tread.
The night comes slowly, and a breeze of roses is blowing.
Let me whisper a ballad, a song of love.
It is dawn. A dove is cooing. Your hair is filled with dew.
Your lips are like a rose to the morning. I shall pick it for myself.

HAVAH NAGILAH

הָבָה נָגִילָה וְנִשְׂמְחָה.
הָבָה נְרַנְּנָה וְנִשְׂמְחָה.
עוּרוּ אַחִים בְּלֵב שָׂמֵחַ.

Come, let us be glad and rejoice.
Let us sing and rejoice. Awake, friends, with a joyful heart.

HAVAH NEITZEI B'MACHOL

הָבָה נֵצֵא בְּמָחוֹל
הָבָה נֵצֵא בִּמְחוֹלוֹת
יַלֵילִי . . .

Come, let us go out and dance.

ITI MIL'VANON

אִתִּי מִלְּבָנוֹן, אִתִּי כַּלָּה תָּבוֹאִי
מִמְּעֹנוֹת אֲרָיוֹת מֵרֹאשׁ שְׂנִיר וְחֶרְמוֹן.

הִנָּךְ יָפָה רַעְיָתִי עֵינַיִךְ כַּיּוֹנִים.
זֶה דוֹדִי, זֶה רֵעִי
בְּנוֹת יְרוּשָׁלַיִם, זֶה דוֹדִי.

With me, from Lebanon, come with me, my bride,
from the dens of the lions, from the tops of Mt. Senir and Mt. Hermon.
(Song of Songs 4:8)

Behold, you are beautiful, my beloved. Your eyes are like doves. *(Song of Songs 4:1)*

This is my beloved. This is my friend. Daughters of Jerusalem, this is my beloved.
(Song of Songs 5:16)

KOL DODI

קוֹל דוֹדִי הִנֵּה־זֶה בָּא,
מְדַלֵּג עַל־הֶהָרִים,
מְקַפֵּץ עַל־הַגְּבָעוֹת.

Hark! My beloved comes leaping over mountains, bounding over the hills.
(Song of Songs 2:8)

LU Y'HI

עוֹד יֵשׁ מִפְרָשׂ לָבָן בָּאֹפֶק
מוּל עָנָן שָׁחוֹר כָּבֵד.
כֹּל שֶׁנְּבַקֵּשׁ, לוּ יְהִי.
וְאִם בַּחַלּוֹנוֹת הָעֶרֶב
אוֹר נֵרוֹת־הֶחָג רוֹעֵד.
כֹּל שֶׁנְּבַקֵּשׁ, לוּ יְהִי.

לוּ יְהִי, לוּ יְהִי, אָנָּא לוּ יְהִי,
כֹּל שֶׁנְּבַקֵּשׁ, לוּ יְהִי.

אִם הַמְבַשֵּׂר עוֹמֵד בַּדֶּלֶת
תֵּן מִלָּה טוֹבָה בְּפִיו.
כֹּל שֶׁנְּבַקֵּשׁ, לוּ יְהִי.
אִם נַפְשְׁךָ לָמוּת שׁוֹאֶלֶת
מַפְרִיחָהּ וּמֵאָסִיף.
כֹּל שֶׁנְּבַקֵּשׁ, לוּ יְהִי.

מַה קּוֹל עֲנוֹת אֲנִי שׁוֹמֵעַ
קוֹל שׁוֹפָר וְקוֹל תֻּפִּים!
כֹּל שֶׁנְּבַקֵּשׁ, לוּ יְהִי.
לוּ תִּשָּׁמַע בְּתוֹךְ כָּל אֵלֶּה
גַּם תְּפִלָּה אַחַת מִפִּי.
כֹּל שֶׁנְּבַקֵּשׁ, לוּ יְהִי.

בְּתוֹךְ שְׁכוּנָה קְטַנָּה מֻצֶּלֶת
בַּיִת קָט עִם גַּג אָדֹם.
כֹּל שֶׁנְּבַקֵּשׁ, לוּ יְהִי.
זֶה סוֹף הַקַּיִץ, סוֹף הַדֶּרֶךְ
תֵּן לָהֶם לָשׁוּב הֲלוֹם.
כֹּל שֶׁנְּבַקֵּשׁ, לוּ יְהִי.

There is still a white sail on the horizon opposite a heavy black cloud.
May all that we ask for come to pass.
And if in the evening windows, the light of holiday candles flickers,
may all that we ask for come to pass.

May all that we ask for come to pass.

If the messenger is standing at the door, may he have a good word.
May all that we ask for come to pass.
If your soul seeks to die, from blossom or from harvest.
May all that we ask for come to pass.

What are these sounds I hear, the sound of the shofar, the sound of the drums?
May all that we ask for come to pass.
Let one prayer from my lips be heard from among all these.
May all that we ask for come to pass.

In a small shady neighborhood, there is a small house with a red roof.
May all that we ask for come to pass.
It is the end of the summer, the end of the road; let them come back here.
May all that we ask for come to pass.

MAH NAVU

מַה־נָּאווּ עַל־הֶהָרִים רַגְלֵי מְבַשֵּׂר,
מַשְׁמִיעַ יְשׁוּעָה מַשְׁמִיעַ שָׁלוֹם.

קוֹל צֹפַיִךְ נָשְׂאוּ קוֹל יַחְדָּו יְרַנֵּנוּ,
כִּי עַיִן בְּעַיִן יִרְאוּ בְּשׁוּב יְיָ צִיּוֹן.

How beautiful on the mountains are the feet of the herald announcing salvation, announcing peace. Hark! Your sentries raise their voices, as one they shout for joy, when every eye shall behold Adonai's return to Zion. *(Isaiah 52:7–8)*

OD YAVO SHALOM ALEINU

עוֹד יָבוֹא שָׁלוֹם עָלֵינוּ וְעַל כֻּלָּם.
סָלָאם, עָלֵינוּ וְעַל כָּל הָעוֹלָם,
סָלָאם, סָלָאם.

Peace will surely come to us, to everyone. Salaam, for us and for the entire world.

SHIR LASHALOM

תְּנוּ לַשֶּׁמֶשׁ לַעֲלוֹת לַבֹּקֶר לְהָאִיר,
הַזַּכָּה שֶׁבַּתְּפִלּוֹת אוֹתָנוּ לֹא תַּחֲזִיר.
מִי אֲשֶׁר כָּבָה נֵרוֹ וּבֶעָפָר נִטְמָן,
בֶּכִי מַר לֹא יָעִירוּ לֹא יַחֲזִירוֹ לְכָאן.

אִישׁ אוֹתָנוּ לֹא יָשִׁיב
מִבּוֹר תַּחְתִּית אָפֵל, כָּאן לֹא יוֹעִילוּ
לֹא שִׂמְחַת הַנִּצָּחוֹן וְלֹא שִׁירֵי הַלֵּל.

לָכֵן רַק שִׁירוּ, שִׁיר לַשָּׁלוֹם
אַל תִּלְחֲשׁוּ תְּפִלָּה
מוּטָב תָּשִׁירוּ, שִׁיר לַשָּׁלוֹם
בִּצְעָקָה גְּדוֹלָה.

תְּנוּ לַשֶּׁמֶשׁ לַחֲדֹר מִבַּעַד לַפְּרָחִים.
אַל תַּבִּיטוּ לְאָחוֹר, הַנִּיחוּ לָהוֹלְכִים.
שְׂאוּ עֵינַיִם בְּתִקְוָה, לֹא דֶּרֶךְ כַּוָּנוֹת
שִׁירוּ שִׁיר לָאַהֲבָה, וְלֹא לַמִּלְחָמוֹת.

אַל תַּגִּידוּ יוֹם יָבוֹא, הָבִיאוּ אֶת הַיּוֹם!
כִּי לֹא חֲלוֹם הוּא
וּבְכָל הַכִּכָּרוֹת הָרִיעוּ לַשָּׁלוֹם!

Let the sun rise, the morning dawn. The purest of prayers will not bring us back. Bitter crying will not awaken or return those whose candle has gone out and who have been buried in the dust. It won't return anyone to us from dark pits. Here neither joy of victory nor songs of praise will be of any use.

So just sing, sing for peace! Don't whisper a prayer. It is far better to sing for peace, in one great shout.

Let the sun shine through the flowers. Don't look back; leave that to pedestrians. Lift up your eyes in hope, not through gun sights. Sing a song of love, not of war!

Don't say the day will come. Bring the day! Because it is not a dream. In every city square cry out for peace.

TZADDIK KATAMAR

צַדִּיק כַּתָּמָר יִפְרָח;
כְּאֶרֶז בַּלְּבָנוֹן יִשְׂגֶּה.

The righteous bloom like a date-palm; they thrive like a cedar in Lebanon.

(Psalm 92:13)

USH'AVTEM MAYIM

וּשְׁאַבְתֶּם מַיִם בְּשָׂשׂוֹן
מִמַּעַיְנֵי הַיְשׁוּעָה.

Joyfully shall you draw water from the wells of salvation. *(Isaiah 12:3)*

Y'RUSHALAYIM (MEI-AL PISGAT HAR HATZOFIM)

מֵעַל פִּסְגַּת הַר הַצּוֹפִים,
שָׁלוֹם לָךְ יְרוּשָׁלַיִם.
מֵעַל פִּסְגַּת הַר הַצּוֹפִים,
אֶשְׁתַּחֲוֶה לָךְ אַפָּיִם.
מֵאָה דוֹרוֹת חָלַמְתִּי עָלַיִךְ,
לִזְכּוֹת לִרְאוֹת בְּאוֹר פָּנָיִךְ.

יְרוּשָׁלַיִם, יְרוּשָׁלַיִם!
הָאִירִי פָּנַיִךְ לִבְנֵךְ!
יְרוּשָׁלַיִם, יְרוּשָׁלַיִם!
מֵחָרְבוֹתַיִךְ אֶבְנֵךְ!

From the peak of Mt. Scopus, shalom, Jerusalem! From the peak of Mt. Scopus, I bow down low before you. A hundred generations I have dreamed of you, dreamed of the privilege to bask in your light. Jerusalem, Jerusalem! Smile on your children! Jerusalem, Jerusalem! Out of your ruins will I rebuild you!

Y'VARECH'CHA ADONAI MITZION

יְבָרֶכְךָ יְיָ מִצִּיּוֹן וּרְאֵה
בְּטוּב יְרוּשָׁלָיִם כֹּל יְמֵי יְמֵי חַיֶּיךָ;
וּרְאֵה־בָנִים לְבָנֶיךָ שָׁלוֹם עַל־יִשְׂרָאֵל.

May God bless you from Zion; may you see Jerusalem's well-being all the days of your life. And may you see children for your children. Shalom for Israel!

(Psalm 128:5-6)

Y'RUSHALAYIM SHEL ZAHAV

אֲוִיר הָרִים צָלוּל כַּיַּיִן וְרֵיחַ אֳרָנִים
נִשָּׂא בְּרוּחַ הָעַרְבַּיִם עִם קוֹל פַּעֲמוֹנִים.
וּבְתַרְדֵּמַת אִילָן וָאֶבֶן שְׁבוּיָה בַּחֲלוֹמָהּ
הָעִיר אֲשֶׁר בָּדָד יוֹשֶׁבֶת וּבְלִבָּהּ חוֹמָה.

יְרוּשָׁלַיִם שֶׁל זָהָב
וְשֶׁל נְחֹשֶׁת וְשֶׁל אוֹר
הֲלֹא לְכָל שִׁירַיִךְ אֲנִי כִּנּוֹר!

אֵיכָה יָבְשׁוּ בּוֹרוֹת הַמַּיִם,
כִּכַּר הַשּׁוּק רֵיקָה.
וְאֵין פּוֹקֵד אֶת הַר הַבַּיִת
בָּעִיר הָעַתִּיקָה.
וּבַמְּעָרוֹת אֲשֶׁר בַּסֶּלַע
מְיַלְלוֹת רוּחוֹת.
וְאֵין יוֹרֵד אֶל יַם הַמֶּלַח
בְּדֶרֶךְ יְרִיחוֹ.

אַךְ בְּבוֹאִי הַיּוֹם לָשִׁיר לָךְ
וְלָךְ לִקְשֹׁר כְּתָרִים,
קָטֹנְתִּי מִצְּעִיר בָּנַיִךְ
וּמֵאַחֲרוֹן הַמְשׁוֹרְרִים.
כִּי שְׁמֵךְ צוֹרֵב אֶת הַשְּׂפָתַיִם
כִּנְשִׁיקַת שָׂרָף.
אִם אֶשְׁכָּחֵךְ יְרוּשָׁלַיִם אֲשֶׁר כֻּלָּהּ זָהָב.

חָזַרְנוּ אֶל בּוֹרוֹת הַמַּיִם לַשּׁוּק וְלַכִּכָּר
שׁוֹפָר קוֹרֵא בְּהַר הַבַּיִת בָּעִיר הָעַתִּיקָה.
וּבַמְּעָרוֹת אֲשֶׁר בַּסֶּלַע
אַלְפֵי שְׁמָשׁוֹת זוֹרְחוֹת
וְשׁוּב נֵרֵד אֶל יַם הַמֶּלַח
בְּדֶרֶךְ יְרִיחוֹ.

The mountain air is clear as wine and the fragrance of pine is carried in the evening breeze with the sound of bells. In the slumber of tree and stone, captive within her dream, is the city which sits deserted, and the wall at its heart.

Jerusalem of gold, of bronze, and of light, am I not a harp for all your songs?

How the cisterns have dried up! The market square is empty. No one attends the Temple Mount in the Old City. And in the caves in the rock winds moan. No one descends to the Dead Sea by way of Jericho.

But when I come today to sing unto you and to bind crowns for you, I become smaller than the youngest of your sons or the least of the poets. For your name burns the lips like the kiss of a seraph if I forget you, O Jerusalem, that is all of gold.

We have returned to the cisterns, to the market and to the square. A shofar calls out on the Temple Mount in the Old City. And in the caves in the rock, thousands of suns shine. We will once again descend to the Dead Sea by way of Jericho.

MEDITATION & HEALING

פִּיּוּטִים

שַׁבָּת

הַבְדָּלָה

שִׁירִים

שִׁירֵי אֶרֶץ יִשְׂרָאֵל

מֶדִיטַצְיָה וְהַחְלָמָה

שִׁירִים לְאֻמִּיִים

OPEN UP OUR EYES

Open up our eyes, teach us how to live
Fill our hearts with joy and all the love You have to give
Gather us in peace
As You lead us to Your Name
And we will know that You are One.

KOL HAN'SHAMAH

כֹּל הַנְּשָׁמָה תְּהַלֵּל יָהּ, הַלְלוּ, הַלְלוּ־יָהּ.

Let all that breathes praise God! Hallelujah!   *(Psalm 150:6)*

EL NA R'FA NA LAH

אֵל נָא רְפָא נָא לָהּ.

God, I pray, heal her.   *(Numbers 12:13)*

NATIONAL HYMNS

HATIKVAH

כָּל עוֹד בַּלֵּבָב פְּנִימָה
נֶפֶשׁ יְהוּדִי הוֹמִיָּה
וּלְפַאֲתֵי מִזְרָח קָדִימָה
עַיִן לְצִיּוֹן צוֹפִיָּה.
עוֹד לֹא אָבְדָה תִּקְוָתֵנוּ
הַתִּקְוָה בַּת שְׁנוֹת אַלְפַּיִם
לִהְיוֹת עַם חָפְשִׁי בְּאַרְצֵנוּ
אֶרֶץ צִיּוֹן וִירוּשָׁלָיִם.

So long as within the inmost heart a Jewish spirit sings, so long as the eye looks
eastward, gazing toward Zion, our hope is not lost — the hope of two thousand years:
to be a free people in our land, the land of Zion and Jerusalem.

STAR SPANGLED BANNER

O say, can you see
By the dawn's early light,
What so proudly we hailed
At the twilight's last gleaming?
Whose broad stripes and bright stars,
Through the perilous fight,
O'er the ramparts we watched
Were so gallantly streaming!

And the rockets' red glare,
The bombs bursting in air,
Gave proof through the night
That our flag was still there!
O say, does that star-spangled banner yet wave
O'er the land of the free, and the home of the brave?

O CANADA

O Canada!
Our home and native land!
True patriot love in all thy sons command.

With glowing hearts we see thee rise,
The True North strong and free!

From far and wide,
O Canada, we stand on guard for thee.

God keep our land glorious and free!
O Canada, we stand on guard for thee.

O Canada, we stand on guard for thee.

GOD BLESS AMERICA

God bless America,
Land that I love.
Stand beside her, and guide her
Thru the night with the light from above.
From the mountains, to the prairies,
To the oceans, white with foam
God bless America,
My home sweet home.

AMERICA THE BEAUTIFUL

O beautiful for spacious skies,
For amber waves of grain,
For purple mountain majesties
Above the fruited plain!

America! America!
God shed His grace on thee,
And crown thy good with brotherhood
From sea to shining sea!

O beautiful for pilgrim feet
Whose stern impassion'd stress
A thoroughfare for freedom beat
Across the wilderness.

America! America!
God mend thine ev'ry flaw,
Confirm thy soul in self-control,
Thy liberty in law.

O beautiful for heroes prov'd
In liberating strife,
Who more than self their country loved,
And mercy more than life.

America! America!
May God thy gold refine
Till all success be nobleness,
And ev'ry gain divine.

O beautiful for patriot dream
That sees beyond the years
Thine alabaster cities gleam
Undimmed by human tears.

America! America!
God shed His grace on thee,
And crown thy good with brotherhood
From sea to shining sea.

ADDITIONAL SONGS

Permissions

Every effort has been made to ascertain the owners of copyrights for the selections used in this volume and to obtain permission to reprint copyrighted passages. The Central Conference of American Rabbis expresses its gratitude for permissions it has received. The Conference will be pleased, in subsequent editions, to correct any inadvertent errors or omissions that may be pointed out.

In addition to the specific acknowledgments listed below, the CCAR extends its thanks to the Jewish Publication Society of America for permission to adapt selected Psalm texts from the *JPS Tanakh: The Holy Scriptures*; to Transcontinental Music Publications for providing many of the song texts in *Mishkan T'filah*, and to the URJ Press for its ongoing guidance and support.

Abrams, Judith Z., for the selections *Adonai, our God, grant us knowledge*; *In the morning*; and several notes.

Acum Ltd: *Al Kol Eileh,* text by Naomi Shemer; *Ani V'atah,* text by Arik Einstein; *Bashanah Haba'ah*, text by Ehud Manor, *Eli Eli (Halichah L'Keisariah*, text by Hannah Senesh; *Erev Shel Shoshanim*, text by Moshe Dor; *Lu Y'hi*, text by Naomi Shemer; *Od Yavo Shalom Aleinu*; *Shir Lashalom*, texts by Moshe Ben-Ari; *Y'rushalayim Shel Zahav*, text by Naomi Shemer; *Yerushalayim (Me'al Pisgat Har Hatzofim),* text by Avigdor Hameiri. Used by permission.

Friedman, Debbie: *L'chi Lach; Miriam's Song; T'filat Haderech; Mi Shebeirach; Im ain ani, El na* by Debbie Friedman. *Birkat Gomeil* by Debbie Friedman and Elyse Frishman, used by permission of the authors.

Holy Cow! Press: Excerpts from *Harvest, Collected Poems and Prayers*, by Ruth Brin, published 1999; used by permission of The Permissions Company Rights Agency for Holy Cow! Press.

Jewish Lights Publishing: Excerpt from *My People's Prayer Book: Traditional Prayers, Modern Commentaries Vol. 7: Shabbat at Home* © 2003 Rabbi Lawrence Hoffman (Jewish Lights Publishing); Excerpts from *My People's Prayer Book: Traditional Prayers: Modern Commentaries Vol. 2: The Amidah* © 1998 Rabbi Lawrence Hoffman. Permission granted by Jewish Lights Publishing, P.O. Box 237, Woodstock, VT 05091 www.jewishlights.com

Jewish Publication Society: Adaptations from *Tanakh: The Holy Scriptures,* © 1985, published by the The Jewish Publication Society with the permission of the publisher, The Jewish Publication Society.

Klepper, Jeffrey: "*Shavua Tov,*" "This Is Very Good," and "Open Up Our Eyes."

Levy, Richard: selections from *As I wrap myself* and several notes.

Malloy, Merrit: "When I die," used by permission.

Random House, Inc.: From *Entering Jewish Prayer* by Reuven Hammer © 1994 by Reuven Hammer. (Schocken Books, a division of Random House, Inc.)

Sol, Adam, for contributing the selection "God can hardly be listening."

Sounds Write Productions, Inc: Text of *Shir Chadash* in the English / Hebrew version by Julie Silver. Used by permission of Sounds Write Productions, Inc.

Steinbaum, Ellen, for "We enter this sanctuary."

Source Citations

Weekday Evening

2 *We are called unto life . . .* Elyse Frishman, adapted

23 *In the morning . . .* Judith Z. Abrams and Elyse Frishman, adapted from *Y. B'rachot 4:1, 296*

23 *Within these walls . . .* Ellen Steinbaum

Weekday Morning

26 *As I wrap myself . . .* Richard Levy

46 *Current scholarship . . .* Peter Knobel

56 *You shall always be praised . . .* Jules Harlow, revised

Weekday T'filah

82 *Let us sanctify Your name on earth . . .* based on translation, and with revised direct biblical citations from *JPS Tanakh*

102 *Adonai, our God, grant us knowledge . . .* Judith Z. Abrams, translated into the Hebrew by Yehiel Hayon

103 *Give us insight . . .* translated by Chaim Stern

Reading the Torah on Weekdays

104 *Translations in the Torah ritual . . .* Chaim Stern, freely adapted

109 *Mi Shebeirach . . .* Debbie Friedman and Drorah Setel

Prayers of Our Community

112 *Source of all being . . .* Elyse Frishman

115 *The custom of announcing . . .* Bernard H. Mehlman

116 *Into our hands . . .* adapted by Chaim Stern in *Gates of Prayer* from a prayer in *Union Prayer Book*

116 *Our hearts are one . . .* Elyse Frishman

Kabbalat Panim for Shabbat

124 *We enter this sanctuary to welcome . . .* Ellen Steinbaum

124 *May the door . . .* Sidney Greenberg

126 *I begin with a prayer of gratitude . . .* Ruth Brin

128 *Y'did nefesh . . .* Elezar ben Moses Azikri

128 *Shabbat Hamalkah . . .* Haim Nachman Bialik

129 *Eileh chamdah libi . . .* Elezar ben Moses Azikri

Shabbat Evening I

168 *Tal . . . dew . . .* Bernard H. Mehlman

Shabbat Morning I

184 *Each of us enters this sanctuary . . .* Robert I. Kahn, adapted by Chaim Stern, adapted

190 *As I wrap myself . . .* Richard Levy

208 *Current scholarship . . .* Peter Knobel

247 *Tal . . . dew . . .* Bernard H. Mehlman

Reading the Torah on Shabbat

362 *Translations in the Torah ritual . . .* Chaim Stern, freely adapted

371 *Mi Shebeirach . . .* Debbie Friedman and Drorah Setel

Prayers of Our Community

376 *Source of all being . . .* Elyse Frishman

379 *The custom of announcing . . .* Bernard H. Mehlman

380 *Into our hands . . .* adapted by Chaim Stern in *Gates of Prayer* from a prayer in *Union Prayer Book*

380 *Our hearts are one . . .* Elyse Frishman

Aleinu and Kaddish

592 *When I die . . .* Merrit Malloy

593 *In nature's ebb and flow . . . Union Prayer Book*

593 *The light of life is a finite flame . . .* Chaim Stern

594 *Why should I wish to see . . .* Walt Whitman

594 *It is a fearful thing to love . . .* Chaim Stern

595 *It is hard to sing of oneness . . .* Richard Levy

595 *Yeish kochavim . . .* Hannah Senesh; translation by Jeff Klepper and Daniel Freelander
596 *Yizkor . . .* Peter Knobel
596 *Our thoughts turn . . .* Robert I. Kahn, adapted

Blessings for the Home and Synagogue

616 *Elijah the Tishbite . . .* Marc Brettler
616 *Miriam han'via, Miriam the prophet . . .* Leila Gal Berner and Arthur Waskow

Songs and Z'mirot

628 *Yigdal . . .* Dayyan Daniel b. Yehuda
629 *Anim z'mirot . . .* Rabbi Yehuda HaChasid of Regensburg, translated by Howard Apothaker
632 *Ein adir . . .* Yisrael b. Shabtai Hapstein
634 *Yom zeh l'Yisrael . . .* Isaac Luria
634 *Ki eshm'rah Shabbat . . .* Abraham ibn Ezra
635 *Shabbat hamalkah . . .* Haim Nachman Bialik
636 *Y'did nefesh . . .* Eleazar ben Moses Azikri
637 *Yah ribon . . .* Israel Najara
641 *Shavua Tov, Have a Good Week . . .* Jeff Klepper and Susan Nanus
642 *Al sh'loshah d'varim . . .* Shimon HaTzaddik
642 *Eileh chamdah libi . . .* Eleazar ben Moses Azikri
643 *Gesher tzar m'od . . .* Nachman of Bratslav
644 *Im ein ani li mi li? . . .* Debbie Friedman
644 *Im tirtzu . . .* Theodor Herzl and Naftali Herz Imber
645 *L'chi lach . . .* Debbie Friedman and Savina Teubal
645 *Lo alecha . . .* R. Tarfon
646 *Miriam's Song . . .* Debbie Friedman
646 *Mitzvah goreret mitzvah . . .* Andy Vogel
648 *Shir chadash . . .* Julie Silver
649 *T'filat haderech . . .* Debbie Friedman
649 *This is very good . . .* Chaim Stern and Jeff Klepper
651 *Al kol eileh . . .* Naomi Shemer
652 *Ani v'atah . . .* Arik Einstein
652 *Bashanah habaah . . .* Ehud Manor
653 *Eli Eli . . .* Hannah Senesh
654 *Erev shel shoshanim . . .* Moshe Dor
654 *Havah nagilah . . .* Moshe Nathonson
656 *Lu y'hi . . .* Naomi Shemer
657 *Od yavo shalom aleinu . . .* Moshe Ben-Ari
658 *Shir lashalom . . .* Yaakov Rotblit
659 *Y'rushalayim (Mei-al Pisgat Har Hatzofim) . . .* Avigdor Hameiri
660 *Y'rushalayim shel zahav . . .* Naomi Shemer
678 *Open Up Our Eyes . . .* Jeff Klepper
678 *El na r'fa na lah . . .* Debbie Friedman
678 *HaTikvah . . .* Naftali Herz Imber
679 *Star Spangled Banner . . .* Francis Scott Key
679 *O Canada . . .* words by Sir Adolphe-Basile Routhier
680 *God Bless America . . .* Irving Berlin
681 *America the Beautiful . . .* Katharine Lee Bates